TWO MEN CALLED ADAM

DOORWAY
PUBLICATIONS
Brockville, Ontario, Canada

two MEN called ADAM

A fresh look at the creation/evolution controversy from a different point of view~

the **THEOLOGICAL.**

Arthur C. Custance

M.A., Ph.D., F.R.A.I., (ret.)

Other Books by the Author

Without Form and Void (1970)
Noah and his Three Sons (1975)
Genesis and Early Man (1975)
Man in Adam and in Christ (1976)
Evolution or Creation (1976)
The Virgin Birth and the Incarnation (1977)
Time and Eternity (1977)
Hidden Things of God's Revelation (1977)
Science and Faith (1978)
The Flood: Local or Global (1979)
The Sovereignty of Grace (1979)
The Mysterious Matter of Mind (1980)
Indexes to the Doorway Papers Series (1980)
The Seed of the Woman (1980)
Journey out of Time (1981)

ISBN 0 - 919857 - 02 - 7
National Library of Canada

P.O. Box 291, Brockville, Ontario, Canada K6V 5V5

AN ACKNOWLEDGEMENT

To **Evelyn May White:**

for whom I owe a great debt to the Lord for sending me the best secretary an author ever had, and the most faithful of all critics.

She took all the dictation at least twice over and typed much of the text at least three times: and then set the copy on a Composer ready for the printer, thus making many last-minute refinements possible.

I am grateful beyond telling.

Note about the documentation:

Every statement made in this volume can be documented, appropriately. In order to simplify the text for the general reader, we decided to avoid the annoyance of the repeated call numbers it entails.

We have also reduced footnotes to a bare minimum for the same reason.

Anyone who is genuinely concerned about any particular statement is invited to write to the author via the publisher's address and he will be happy to provide the source of his information.

TABLE OF CONTENTS

3 PROPOSITIONS

An established fact is as sacred as a revealed truth.

Nothing quite equals the ignorance of the average Scientist about Theology except perhaps the ignorance of most Theologians about matters of Science.

"Theology was once known as the Queen of the Sciences. If Science as the 'Servant of Humanity' is to be sure of its direction, the Queen needs to be either reinstated or replaced . . . . THE THRONE IS AT PRESENT VACANT."

A. V. Hill, ***Presidential Address*** **before the British Association in 1954.**

Why is it that the theologians are just as unwilling to incorporate the data *of science into their theology as the scientists are of incorporating the* data *of theology into their science? These data in* both *cases ultimately rest upon foundations of a similar nature, namely, on the logical extension of the implications of premises which have been accepted by faith.*

INTRODUCTION

In the present creation/evolution controversy the basic issue is the origin and nature of Adam. Was he only a little removed from the apes but with the benefit of a soul, appearing on the scene millions of years ago? Or was he a creature of God's making coming directly into being by a divine act, placed in a Garden of Eden only a few thousand years ago? And what of Eve? Was she a kind of prehistoric "Lucy" three million years old, or was she formed out of a historical Adam by a unique act of divine surgery as recorded in Genesis so matter-of-factly?

What has been overlooked in all this controversy, and what complicates it immensely, is the fact that in the Bible we have not merely *one* man called Adam to account for, but *two* (1 Cor.15:45). And these two men, the Adam of Eden and the Adam of Bethlehem, stand in direct apposition to one another, each being a prototype and a representative of the other and of true Man. Whatever we can say with certainty about the Last Adam must be assumed for the First Adam as originally formed – whether by creation or by evolution.

Examination of the implications of the biblical data (and this data is far more revealing than is generally recognized) shows that the creation of Adam *and* Eve exactly as Scripture sets it forth, is the *only* view that really makes sense of the relationship between these two Adams and of the biblical Plan of Redemption. Evolution makes a shambles of the Plan of Redemption as the Church of God has understood it and preached it for almost 2000 years.

It is the exploration in depth of this relationship between the First and the Last Adam that forms the subject of this volume. It demonstrates that in this controversy concerning the origin of man's body, the biological data are only part of the issue: it is the theological data that must now be addressed.

What the Lord Jesus Christ accomplished on Calvary cannot really be understood without first grasping the difference between our physical death and that of an animal, and between the death of Jesus Christ and that of us men for whom He died. These three kinds of physical death are entirely different in certain fundamental essentials. It is quite beyond the ability of evolutionary theory to account for these physiological differences.

Chapter 1

THE PROBLEM STATED

Can Evolutionary Theory and Christian Theology Be Reconciled?

When Darwin in 1859 published his *Origin of Species*, it does not seem to have troubled the theologians too greatly since he was careful not to make the evolution of man an explicit corollary. But when in 1871 he published his *Descent of Man*, the true significance of the threat to Christian Faith became much more apparent.

In the first place, it was obvious that the concept of the origin of man by evolution from some ape-like creature ran counter to the concept of man as a special creation of God made in the Creator's own image and possessing from the first a high intelligence and moral freedom.

In the second place, it was obvious that the evolutionary view gave man an antiquity vastly in excess of the mere 6000 years traditionally held on the basis of a strictly biblical chronology.

But there were other consequences which would have to be faced in due time.

Evolution makes it impossible to establish the precise antithetical relationship between the Last Adam and the First Adam, so essential to his role as Saviour-substitute for man. Was each truly a reflection of the other if the First Adam was as much ape as he was man, a barely human figure lurking in fear of life and limb in some dark cold cave and surviving by the barest margin for thousands of years until intelligent enough to secure a measure of superior self-sufficiency? Could such a half-human figure be a prototype of the Lord Jesus Christ?

Then there is the theological requirement that Adam's constitution possess a contingently immortal nature, a physical body not inherently subject to death yet quite capable of experiencing death as the penalty of disobedience. Such a physical being could not be derived from any one of the then existing primates for whom it is almost certain that death was "programmed" to occur after a certain species-specific span of life had passed.

There are other problems, too. Clearly Eve was formed out of Adam. There is no other species of which the female was, as it were, born out of the male.

Now it is clear from subsequent events after the publication of *The Origin* and *The Descent of Man* that the full theological implications of some of these difficulties were not recognized at the time by evangelical theologians. There were, however, two specific problems that they did indeed recognize from the very first and these were felt to be most disturbing.

These two problems were: (1) the enormous antiquity of man which apparently vastly exceeded the allotted biblical starting date of 4004 BC; and (2) the story of man's introduction into the world by divine intervention in the form of a direct creative act, not out of a lower order of animal life but out of the dust of the earth and in the Creator's image.

Now both these problems were resolved by re-interpreting the biblical chronology, and by re-interpreting the concept of evolution to mean in Adam's case that this was the divine mode of man's "creation".

It is my object to analyze the factors in the new theory

of origins which from a theological point of view were most serious, and to show how, historically, the seriousness was not recognized by those who should in fact have been the first to sound the alarm. It will be useful to review very briefly as a kind of introduction, what happened at Princeton Theological Seminary which at the time was a centre of evangelical theology marked by the appearance of a succession of perhaps the greatest Systematic Theologians America was to produce for many years. What is surprising is the speed with which capitulation to Darwinism took place.

Indeed, by 1900 capitulation to Darwinism at Princeton on these two critical issues was virtually complete. The other problems that Darwinism generated for those who placed their faith in the Word of God crystallized much more slowly. But what had already been surrendered by 1900 was enough to ensure that many other theological seminaries and colleges soon took a more liberal view of the matter of human origins and increasingly departed from the biblical position held by their founding fathers.

While this sad surrender to evolutionism was gathering momentum in the final quarter of the last century, Princeton was blessed with a succession of evangelical giants: Charles Hodge (1797-1878) the leading American theologian of the nineteenth century; his son Alexander A. Hodge (1823-1886) who followed him in the Chair of Systematics; William H. Green (1825-1896) who became Professor of Oriental and Old Testament literature and wrote a paper on *Primeval Chronology* that had at that time (and still has) a profound influence on the whole question of the antiquity of Adam (as we shall see in Chapter 18); and Benjamin B. Warfield (1851-1921) who was a product of Princeton and in due course became Professor of didactic and polemical theology at his Alma Mater, succeeding A. A. Hodge.

Charles Hodge at first took the position that evolution was a highly speculative theory and far from being proved.

He never for a moment conceded that man's *spirit* was evolved and he had serious doubts if man's body could have been either. But with the publication of Sir Charles Lyell's *Antiquity of Man* in 1863,* he began to have doubts whether the biblical chronology which placed the creation of Adam barely 6000 years ago could any longer be defended.

Like his father, Alexander Hodge was much impressed by the manifest "devoutness" of Lyell and as a result his conclusion met with considerable sympathy in Hodge's mind. While he admitted the evidence for a vast antiquity of the earth and that man may have been introduced upon it much earlier than the Genesis record seemed to allow, he wrote: "In any event, it can prove nothing as to the relation of Adam to the race, but only that he was created longer ago than we suppose". This was the thin edge of the wedge.

The idea of the vast antiquity of man had also troubled William H. Green, and in a work designed to defend the Genesis story against the violent attacks of Bishop Colenso, Green published in 1863 his book *The Pentateuch Vindiccated from the Aspersions of Bishop Colenso* in which he argued that the biblical chronology could not be attacked on the grounds that it did not accord with the antiquity of man established from *geological* evidence because the Bible did not actually provide an unbroken chronology in any case. The genealogies by which the Ussher chronology had been constructed appeared to Green to possess many gaps, leaving the date of Adam's creation quite uncertain. The traditional date of 4004 BC could now therefore be abandoned on biblical grounds!

---

* It is worthy of note that Lyell had to publish his *Antiquity of Man* in 1863 before Darwin felt it safe to publish his *Descent of Man* in 1871. The first made room for the second, and the second made explicit what was implied in the first.

The younger Hodge towards the end of his life described the powerful effect Green's book had on his father. In his own words: "I can well remember my father walking up and down in his study when he heard (about it) and saying, 'What a relief it is to me that he should have said that'." It appeared to resolve any conflict between the Bible and geology.

It does not seem that the younger Hodge recognized the full import of this "escape hatch". Green had written, "the time between the creation of Adam and ourselves might have been, for all we know from the Bible to the contrary, much longer than it seems". To Alexander, this only meant that Adam must be placed further back in time – not that Adam was any different in nature from the traditional view of him or that his subsequent history needed any revision. Adam was still a unique creation.

Warfield, who succeeded him, accepted this judgment – again, without apparently realizing what the consequences of such a vast antiquity could be for the first three chapters of Genesis. He, too, accepted Adam as somehow a true "creation" but his origin could be placed much further back in geological time for all it mattered. Thus he wrote in 1911, "The question of the antiquity of man has of itself no theological significance. It is to theology, as such, a matter of entire indifference how long man has existed on earth".

Throughout this debate it must be borne in mind that the issue revolved around the origin of Adam's body. The origin of his soul as a direct creation of God was not for them in question.

Neither the two Hodges, nor Warfield, nor of course Green, was aware that this concession was a very serious one. Yet such a concession plays havoc with the *setting* of the story of the Fall, and therefore with the crucial connection that the New Testament assumes between the First and the Last Adam.

In due course, while holding firmly to the creation of

man's *soul*, they all came to accept the idea of evolution as God's method of producing Adam's *body,* Warfield assuring his readers that evolution was not a substitute for creation but could "supply a theory of the method of divine providence". Man's body fitted readily into the evolutionary chain of life, being the direct result not of outright creation but of millions of years of imperceptible changes from amoeba to man. No supernatural intervention was needed here, nor any literal "divine surgery" for the formation of Eve out of Adam.

It is interesting that while these great defenders of evangelical doctrine were making such tentative concessions to evolution, the liberal theologians were rejecting it – though admittedly for different reasons! And while the evangelicals were increasingly giving support to Darwinism as not incompatible with the Faith they had so ably defended, Darwin himself was steadily surrendering whatever of Christian faith he once had! As we shall see in the final chapter the consequences of admitting the thin edge of the evolutionary wedge were to prove disastrous – not only for Princeton Theological Seminary as a centre of Evangelical Theology, but for many other seminaries on the American continent and in other English speaking countries.

In one instance it took only a single generation to pass from a truly evangelical stance to outright and militant atheism. I have in mind the history of the justly famous Augustus H. Strong (1836-1921) whose *Systematic Theology* is a monumental work of reference which has, since the first edition in 1907, been reprinted at least 29 times. Like his contemporaries he first allowed that Adam's body, but not his spirit, could have been derived by evolution. As he put it, "We concede that man has a brute ancestry".

Then in his less well-known *Christ in Creation*, he frankly admitted that as man received his body thus, there was really no reason why we might not logically admit that this was how he also received his spirit, since this *was* God's method of "creation" in any case.

When he had completed the manuscript for *Christ in Creation*, he requested his son Charles (1863-1940) to proof-read it for him. Although the whole family seems to have shared the father's evangelical faith, it appears that Charles was perceptive enough to see the inconsistency in his father's reasoning. In making such a concession to evolution while at the same time pretending to hold a truly evangelical position, Charles felt his father was compromising himself and misleading his readers.

The end result was that he turned completely against what he was proof-reading, abandoning his position as a Christian, and becoming one of the most ardent and militant proponents of atheism until the day of his death.

But this is not the end of the story, sadly enough. For the Strongs were well-to-do and moved in wealthy circles. Among their friends were such men as John D. Rockefeller, Chauncey Depew, and Andrew Carnegie. The first made his fortune in oil, the second in the railway business, and the third in steel. And each of them was totally ruthless in his business tactics, assuring their critics that they were only acting according to evolutionary principles which were God's methods in nature and therefore good for the species as a whole how ever hard on the individual. They must have received no little comfort from the fact their evangelical friend and scholar, A. H. Strong, could be depended upon to support them in their philosophy.

Why did men of such learning, such dedication to Scripture, such a profound grasp of biblical theology, so easily allow evolutionary philosophy to poison their own thinking? To a man, save only for Augustus H. Strong, they held resolutely to the supernatural creation of the *soul* of the first man. Why, then, did they yield so quickly to an entirely materialistic process of evolution for Adam's *body*, in view of their commitment to the Word of God which is so clear on the matter?

Did they not recognize that the implacable offensive of

evolutionary philosophy would never stop until *all* supernaturalism was abolished entirely? It was only a matter of time till the evolutionists would insist with equal dogmatism that man's "soul" was originated in the same was as his body. Even A. H. Strong could anticipate this. As Kirtley F. Mather quite recently put it, "The spiritual aspects of the life of man are just as surely a product of the processes called evolution as are his brain and his nervous system". Granted this logical extension, it is absurd to talk of the "saving of the soul for eternity". If the *soul* is a mere epiphenomenon of the human body as *consciousness* is an epiphenomenon of the animal body, then soul, like consciousness, must perish with the body – and what evolutionist would be either willing or logically competent to defend such a supernatural event as resurrection of the body? The Greeks thought the idea utterly ridiculous because they held that man was distinctively a spiritual being and the body a prison, from which the spirit longed to be free.

Why do evangelicals fail to recognize the human body as a fundamental complement of the human spirit? Man is not a spiritual being who happens to have a body which he can do without quite easily on the other side of the grave. His body was created to serve as a *permanent* house for his spirit. It was not the divine intention that the body and the spirit should be rent apart. Even A. A. Hodge frankly recognized that Adam's body was potentially immortal at first, and that if he had never sinned he would never have experienced physical death. Unless this is true, death could never have been threatened as a penalty for disobedience.

To derive a *mortal* body out of an *immortal* one is clearly within the competence of evolutionary theory, since the immortality of the lowly amoeba has not been retained in its supposed evolutionary descendants. It has been lost. But to derive an *immortal* body out of a *mortal* one is quite another thing. There is no precedent in nature (with the possible exception of cancer cells). Such a heritage as the derivation of an immortal body out of some form of apelike ancestry as man is presumed to have had, would be so exceptional as to be virtually miraculous and tantamount to an act of creation.

From the theological point of view, an evolutionary derivation of the human body is totally unacceptable. For the Plan of Redemption hinges upon the relationship between two human beings, the First Adam and the Last Adam, the Adam of Genesis and the Lord Jesus Christ of the Gospels whom Paul declares to have been the First Adam's counterpart: two persons who stand in direct apposition both as to body and as to spirit.

Each precisely reflects the other in terms of their human constitution. In both body and soul the Last Adam must truly match the First Adam in every sense if substitution is to be effective. The *first* human being (whatever we conceive his outward form to have been and how ever little removed from the apes) must be every whit as redeemable as the very last human being yet to be born within the framework of this present world's natural order before the Creator pulls down the curtain upon it.

The first man must come as effectively under the Redeemer's umbrella as the last man yet to be born will have to. Thus the first man Adam and all his descendants from the first to the last – i.e., all "in Adam" – must be essentially indistinguishable from each other. *Evolutionary progress* **within** *the human species cannot be reconciled with a Plan of Redemption which depends upon the death of One who appeared so late in the chain. And He appeared late indeed if hundreds of thousands or even millions of years intervened. If the evolution of man is true, this Redeemer, in his far more advanced state of evolution, would no longer represent those who had appeared in a much more primitive state at the beginning of the line.*

Now man, as a creation of God, is never viewed in Scripture as essentially a spiritual being who just happens to have a body. Man is a hyphenate being, a unified body-spirit entity. Without the body the spirit is not a *person* and without the spirit the body is a mere corpse. As much is said in the New Testament about the destiny of the body as is said of the destiny of the spirit, and more is said about the form the body will take than is said about the form the spirit will take. We shall certainly not be mere ghosts!

When the Lord became Man, He did so by assuming a prepared body. When He rose from the dead, He rose in his *own* body. When He returns, He will return in his *own* body; and meanwhile He is as a MAN in heaven because He is a human spirit in a human body. This in no way challenges the fact that in becoming Man He never ceased to be what He has always been: God with God and equal with the Father. In this there is no contradiction. It is rather that in Him now dwells all the fullness of the Godhead BODILY, which in his Person has added a new dimension. Thus it is proper to watch for the glorious return of the Son of MAN, for in his resurrected body so will He appear.

I do not wish to pursue this further at the present moment since it is dealt with later. I wish only to point out that the incarnate Lord assumed a body that was truly and unequivocally human. This body formed a perfect vehicle for the human spirit which He created for himself, as He himself had created one for Adam in order to provide a perfect vehicle for *his* spirit. It is inconceivable that the Lord could be incarnate in a body subject to steady decay, increasing senility, and final total collapse due to old age. If in an immortal body He was truly Man, then true manhood in the First Adam as created equally immortal by nature, demanded a body no less free from the burdens and limitations of physical senility and death due to old age.

By such a means, the pre-existent Lord of Glory condescended to engage for himself a human form that truly reflected Adam as he came from the hand of God. In the humanity of Christ Jesus we therefore see restored to our view the form of the very first human being both as to his body and as to the spirit which animated it in the very beginning. The break between ape and man was absolute.

There were no half-way houses in the line of Adam's seed between the animal and the human world. By divine intervention the body of the *first man* was created uniquely, thus forming a total discontinuity in the great chain of being, a physical organism perfectly suited to be in due course a housing for the Creator himself. This was a new thing: a thing apart though with many shared functions which involved homologies in design because the world into which man was thus introduced was a world in which animals also were designed to function. Both lived in the same environment: both were embodied.

This new creature, unlike all that preceded, was unique in many ways, but in two ways above all – that he was both capable of falling and equally capable of being redeemed; and that he could die but was under no *necessity* of ever doing so. Moreover, his redemption was possible only because he was constituted with a human spirit of God's creation and a human body also of God's creation, both of which clearly involved supernatural intervention – and this was true in *both* Adams, the First and the Last.

In the Person of Jesus Christ as Man, *redemption was secured by death, the death of a human spirit and the death of a human body*. But neither kind of death was natural in the sense of being inevitable for inherent reasons. Both deaths were experienced by an act of will.

But the possession by the First Adam of a body derived from some non-human primate ancestor would have made death inevitable for him. Death per se would not then be the penalty for disobedience but merely *premature* death. By the same token, if the Redeemer was provided with a body human only in this evolutionary sense, death for Him could not have been substitutionary but would have been likewise merely premature.

Equally crucial to the Plan of Salvation, this Adamic body, which *could* experience death but was under no necessity of doing so, in the Second Adam had to be truly representative of the body of the First Adam as created. Without this correspondence there could be no substitution.

These very precise requirements of the nature of Adam's body entirely preclude an animal origin. If these requirements are not fulfilled, no redemption of man is possible. The logic of this is compelling and leaves no room for an evolutionary origin of man.

Man must be supplied from the first with such a body in order that, when he fell, the Redeemer could undo the Fall by becoming Man himself, *because our Substitute must be both truly a human being and yet truly under no necessity of dying*. Otherwise his death was not substitutionary *for man*.

So it has to be asked, At what point in this evolutionary "Chain of Being" did true man emerge as identifiably human in each of these particular respects, in order that the Redeemer by assuming such a body and such a spirit could act as a true representative?

When we bear in mind that this Redeemer was himself very God, the Creator of the Universe, is it conceivable that such a One could assume a form even less human in appearance than, let us say, Neanderthal man? In the quaint words of the King James Version, *his* temple must be "magnifical".

In actual fact we ourselves almost certainly compare very unfavourably with the First Adam. Such was the magnificence of his body that he lived for almost a thousand years despite the fact that he was already suffering the catastrophic effects of the Fall which must steadily have been robbing him of most of his pristine glory.

No million-year-old, tiny-brained, half-ape creature, such as the famous "Lucy", could thus have been an Adam (or an Eve!), especially in view of the fact that Adam's descendants within three or four generations had already developed city life, agriculture, metallurgy, and the musical arts with wind and stringed instruments (Gen.4:21).

As a faithful representative of the First Adam, the Second Adam lived among us a Man – indeed pre-eminently so, as his title "Son of Man" indicates. And He left a transforming stamp upon the history of mankind because of his stature. Even physically He awed those about Him.

We know that He created man in his own image. He clearly did this in order that He himself might one day come among *us* in the image of Man without doing any violence whatever to his own divine nature nor even demeaning his own pre-existent glory, a glory to which He returned. And when, by ascension, He did return to that glory, He did not lay aside his human body as though it were unworthy. He took it with Him. He is, indeed, embodied as Man for ever, and as He ascended into heaven, so will He return in like manner. The human body is a unique vehicle and holds within itself the promise of an almost incredible glory. It is a special creation of God, not a mere by-product of a blind process as we are being asked to believe.

We easily fall into the trap of relegating the body to a very second-rate position in the scheme of things. As such it seems of little consequence whether its origin was by evolution or creation. We are told that the saving of the soul by the regeneration of the *spirit* is what counts. The redemption of the *body* seems much less important, as though the possession of a sub-human body would serve just as well. Yet we are to continue as a body/spirit entity throughout eternity which includes not only a new heaven but a new *earth*. An earth which is never to grow old (Isa.66:22) would seem to demand a body that will never grow old either!

It is, in fact, the redemption of our *own* body that, as an article of Faith, distinguishes the Christian position from that of every other religious system. It was this kind of resurrection which proved so incredible to Paul's Greek hearers at Athens. "Who ever heard of such a thing?" they asked.* Even today it would seem that many Christians have not heard of such a thing either . . . .

It is this that I wish to address in the 17 chapters which follow. It is my sincere hope that men far better theologically informed than myself will take up the issue and carry the matter much further than I am able to do. When this has been done, I predict we shall suddenly recognize that in the current creation/evolution controversy, there has been an important "missing dimension". That dimension represents, according to Scripture, the other half of the Faith: the redemption of the body as essential to the salvation of the soul. The body is very important, but evolution denigrates it, making it merely an animal body whose destiny is not resurrection but dust.

It may well be objected that my approach is far too literalistic, that I am leaning much too heavily on logical analysis of the actual words themselves and thus destroying the spirit of them.

But I want to show that, unless we have evidence from Scripture itself to the contrary, we should take the wording at its face value as our starting point. Otherwise we ought to abandon all pretence that we really are making the Bible the touchstone of our thinking in all matters where revelation plays an essential role.

---

* So totally foreign was the idea to the Greeks that they mistook Paul's term for *resurrection (anastasis)* for the name of a new deity and asked what new god he was speaking about (Acts 17:32)! Plato considered that the body imprisoned the spirit, and therefore that death was the liberation of it. Paul, on the other hand, viewed the body as essential for the spirit to express itself, so that he saw disembodiment as effectively a crippling of it. Thus, for Plato embodiment was a *penalty*: whereas for Paul *dis*-embodiment is the penalty. The two positions, the Greek and the Christian, are diametrically opposed.

We cannot form a *biblical* theology with any claim to rational coherence which justifies the term *biblical* if we adopt the practice of spiritualizing the text or allegorizing it or reading it as poetry whenever it embarasses us, unless the offending words are actually so treated in the original.

In Scripture parables are always identified as such, poetry is always set forth as poetry (as the Psalms are), and allegories and terms of common parlance like "the rising of the sun" are easily identifiable as to their obvious intent. The first chapters of Genesis are not written as poetry, the sole exception being Genesis 4:23 which is set forth in the original in exactly the same way the text of the Psalms is set forth – a telling indication that the rest of the text is NOT poetry.

So I make no apology for taking the words of Scripture seriously: and after studying this wonderful Book for over fifty years I am convinced that anyone who does read it in this spirit will never fail to marvel at the precision with which the truth is spelled out and hedged against error. Despite the problems of translation here and there (problems which sometimes arise where interpretation is in question), I do not believe that any of the passages upon which my thesis is built involves significant uncertainties of this kind.

As a example of what I mean, some may object to taking the story of Eve's formation out of Adam at its face value. But if – for the sake of discussion – they will tentatively allow its *possibility,* they will, I think, soon see what is far more important, namely, its very *necessity*, if the rationale of the Plan of Redemption is to be preserved. There are many things that current biblical theology has neglected to address because of a failure to attach to the wording of Scripture the seriousness it warrants.

We have long enjoyed a most precise and highly developed "theology of the spirit". It is now time to produce a balancing "theology of the body". When this is done, I think the fundamental issues in the current dispute will be much more apparent. For by conceding to evolution the origin of man's *body*, even though insisting on the divine origin of his spirit by direct creation, we have effectively destroyed *the manhood of man.*

Chapter 2

EMBODIMENT

No Body = 'Nobody'

A justly famous paleontologist, one of America's most informed protagonists for the evolution of man, wrote a few years ago: "There was no anticipation of man's coming. He responds to no plan and fulfills no supernal purpose. He is a unique product of a long unconscious, impersonal, material process . . . . that did not have him in mind. He was not planned".

So thought Professor George Gaylord Simpson. But Simpson was wrong. The appearance of the human body upon this world scene was no accident. Scripture tells us that man was very deliberately planned and created in God's image after what amounts to a divine conference in which God said, "'Let us make man in our own image after our likeness: and let them have dominion over the fish of the sea, and over the fowl of the air, and over the cattle, and over all the earth, and over every creeping thing that creepeth upon the earth'. So God created Adam in his own image, in the image of God created he him" (Gen.1:26, 27). Thus the biblical view of the introduction of man makes him very much the result of a plan.

Bertrand Russell, like many other notable individuals of this age, shared Simpson's dismal view of life and wrote quite as eloquently about the *destiny* of man as Simpson did about his *origin:* "All the labours of the ages, all the devotion, all the inspiration, all the noonday brightness of human genius, are destined to extinction in the vast death of the solar system . . . . No philosophy which rejects these (certainties) can hope to stand".

Such is the logical conclusion of a forthright and consistent materialism which sees man as merely a physical phenomenon among the millions of other physical phenomena which have no more and no less significance in the scheme of things than man. All are by-products of pure accident.

The Universe itself will continue to cool as it has done for billions of years until it dies from loss of useable energy and passes with all its contents into oblivion, unremembered in the unthinkable darkness of the total absence of all consciousness. It will go out, like a match struck for a brief moment that flares and dies as though it had never been. Only silence and cold will remain.

This prospect was accepted without qualification by the Nobel Prize Winner and world renowned French scientist, Jacques Monod, who quite recently stated the case in poignant terms: "Man knows at last that he is alone in the universe's unfeeling immensity, out of which he emerged only by chance . . . . Meanwhile he is left with nothing but an anxious quest in a frozen universe of solitude".

The pathetic sadness and loneliness and pointlessness of human existence lies at the end of this philosophical trail. It is a philosophy of despair. Such is any evolutionary world-view when it is projected with complete consistency to its logical conclusion.

Gresham Machen observed, correctly, that the validity of any system of thought is best evaluated by pursuing the logic of it relentlessly to its ultimate conclusion. The conclusion expressed by these profound scholars speaks for itself. Such candid remarks made by highly informed and intelligent men point up the unhealthiness of any philosophy which renders the individual's life totally insignificant. In the evolutionary world-view, apart from his functions or skills, the individual has no personal worth. He is effectively cancelled out as a person.

Such a tragic view stems from the reduction of man to a mere body, a *thing*, a physical organism essentially nothing more than a machine which will soon wear out to be thrown on the scrap heap.

In an odd way Christians have contributed to this denigration of the whole man. What we have done is to place so much emphasis upon the spiritual welfare of man and so little upon the importance of the body that we have emasculated man. We have tried to make him essentially "angelic", a spiritual being who, however, just happens to have a body, a body which we shall be only too happy to slough off. This has had the effect of divorcing two things, body and spirit, which should never even have been separated. To many people it has ceased to be of much importance where the body came from. The origin of his body was left to those who scarcely believed that man had a spiritual side to his nature. We have largely surrendered "to the enemy" all concern for the body, so that effectively, by a joint effort we have annihilated man as man.

We have failed to preserve as part of our Faith any frank acknowledgement of the enormous importance attached, from Genesis to Revelation, to the possession of a *body*. Not just a body of any kind, since all animals have a body and so do plants. No, not just the possession of a body, but the possession of a HUMAN body, a unique house for a unique spirit both of which are of God's creating.

This body is fully one half of our identity as a person. The world was formed in the first place for its very existence and continuance, as Isaiah 45:18 makes clear. "Thus saith the Lord that created the heavens: God himself formed the earth and appointed it; he hath established it, he created it not in vain, *he formed it to be inhabited*". It is no wonder that the astronauts, viewing the earth against the blackness of outer space, saw it as a gem and were deeply moved by it. It seemed to them so beautiful, as home always seems to be when viewed from afar, whatever it may be when examined more closely.

The Medieval theologian, Hugo St. Victor (1096-1141) described the close inter-relatedness of things in a characteristically succinct manner thus: "The spirit was created for God's sake, the body for the spirit's sake, and the world for the body's sake: so that the spirit might be subject to God, the body to the spirit, and the world to the body". Now *there* is a framework for a philosophy of meaning! The body forms the intermediary between God and the physical world through man's spirit.

To complete the sequence, he might have added in the light of what we now understand, "and the Universe for the world" – for such it begins to appear. Revelation 4:11 furnishes the reason why: "For Thou hast created all things, and for thy pleasure they are and were created".

This is even more pointed when it is realized that the phrase "all things" means not merely "everything" in common parlance but more specifically *the Universe*, since this is what the Greek (*ta panta*) signifies. The Universe, and man for whom it was created, was created for his pleasure and still exists for his pleasure – in spite of what man is doing to it.

This present order, however, is temporary. It is to be replaced in due course by a new Universe, not merely by a new heavens only but by a new EARTH also! And this replacement is to be permanent. Isaiah 65:17 and 66:22 as well as Revelation 21:1 give us every assurance of the fact.

If the essence of man is not merely in his spiritual nature but in his physical nature as well, the continuance of man's body is made meaningful by the promise of a new heavens *and* a new earth. In harmony with this promise we are assured that there is reserved for us the prospect of a new "house" for our spirit, a new body which is to be "eternal in the heavens" (2 Cor.5:1).

We have so spiritualized this new Universe which is to be our eternal home that we see man's future as essentially a spiritual one. The spiritual becomes of overriding concern. And yet Colossians 2:17 tells us that we should look upon this present world as a mere shadow of the ageless Universe which is to replace it. And this ageless Universe includes something called a new *earth* with a formal structure such as will accord with our new bodies. We shall not be ghosts flitting from cloud to cloud: in contrast to the angels we shall be *real* people with bodies as real as his resurrected body (Phil.3:21). And He categorically denied that his body was a mere ghost of one.

But why is a body so necessary for man? Angels do very well without them, so why can't we?

There are a number of reasons why man has a body, though it is obvious enough that God can create angels whose existence is just as real despite lack of embodiment. So, allowing that God did not *need* to create man in such a way that he had to have a body, why did He create man thus? Let me suggest five reasons.

(1) One of the most obvious reasons is that *God created a Universe which is physical*. He must have had a reason for doing so and I suppose the best reason has to be simply: Because it suited his purpose. Granted this, we might go one step further and say that He had from the beginning every intention of putting "someone in charge" of it, not in charge of the earth only but ultimately in charge of the whole thing. That someone was man, starting with responsibility for the earth (Gen.1:28). But how could we "take charge" without a body? It does not seem that we could.

Now even though God equipped man with a body for this reason, a body having hands and eyes and ears and feet and a brain and so forth, there was still no guarantee that man would take charge successfully – because he was given freedom to obey or to disobey his mandate. But at least it seems clear that to exercise any kind of dominion over a physical world we had to have these physical appurtenances. Even angels need some temporary embodiment when they are called upon to act on the physical world: not always perhaps, but certainly on occasion – as we shall see later.

Thus, to put the case in a nutshell, man can exercise his will upon the physical order of things only because he *has* these physical properties, i.e., because he has a physical body which is extraordinarily well designed for the express purpose of "taking charge". It is obvious that I can will to move my hand, and it moves. We don't really know how this comes about but it does respond to my willing, and with this ability, we can manipulate material things with a remarkable degree of success – for good or ill. The body is thus a mediator between the *will* and the *world*.

For example, I can't simply will the hands of the clock to move forwards one hour in the spring or backwards one hour in the fall, but I can cause it to happen by making my feet carry me to the clock and making my hand reach up and change the setting to Daylight Saving Time. Indeed, *we can act on the physical world* – even to the extent of beginning the conquest of space – *only because we have a body as our own personal effective instrument*. That is one very good reason why we are given bodies.

(2) But there is a second reason. In order to carry forwards this government of the world, and I suspect of the whole Universe in due course or at least of the New Universe, we must somehow multiply our numbers. There are only two ways by which this increase can be achieved: by direct creation or by procreation.

Let us suppose it is done by creation. *Angels* are individual creations. Each one, as Thomas Aquinas astutely observed, is a separate creation and therefore is a separate species. Each one stands entirely on its own. There are no relationships between them. No angel is father to another angel. They do not multiply by marrying and bearing offspring (Mark 12:25). Thus "to fill the jobs available" in governing the earth, God would have had to create individuals as the tasks developed in order to fill "the positions available".

But now a new problem arises. In the divine plan, these agents, whether pure spirits or embodied, were created with a measure of moral freedom. That angels are not robots is clear from the fact that they do indeed have some measure of freedom, freedom at least to hold contrary opinions.* Furthermore, a number of the angels were disobedient and fell (Jude 6) and these are to be punished for their failure. But then the question arises, If they can be punished, can they also be redeemed?

As far as we can see, they cannot be redeemed. Why not? Well, if we take the Word of God as our guide in all matters of salvation of either men or angels, redemption requires a substitute redeemer, and the redeemer has to be one in form and nature with the subjects to be redeemed. But if each angel is a separate species by the very fact of creation, it would seem that not one redeemer but thousands or perhaps even millions of redeemers would be required, *one for each angel.*

There can be no "first Angel" whose fall involved them all so that they could all be incorporated and redeemed as members of a single family. Since they did not arise by multiplication from a single "father Angel" corresponding to a single "father Adam", the Plan of Redemption which

---

* The evolutionist says that one can have a brain without a mind, and this is certainly true. The brain of Einstein has been preserved in a container since his death, though the rest of his body "lies a mouldin' in the grave". But they then add: "but not a mind without a brain" – and this is certainly *not* true. In this passage (Dan.10:21) the angel, though bodiless, clearly "had a mind of his own". The passage is quite unequivocal about this!

works for man cannot be made to work on the same principle for angels. As the prophet Malachi wrote, "Have we not *one* father?" (2:10) and as Paul re-affirmed in Acts 17: 26, "God hath made out of *one* all men for to dwell on the earth". A *single* redeemer who stands as a second Adam can act as a redeemer of the *whole* species of man. But no such situation is possible with respect to the angels.

The very essence of the Plan of Redemption, substitute of one Head for another, can only be applied to the single species, *Homo sapiens*. To redeem the angels which do not constitute a family at all, would involve as many redeemers as there are fallen angels. This is an inconceivable thing.

Thus, if the appointed government of the material Universe was to be formed of free moral agents who were therefore fall-able* and consequently in need of a redeemer, the only conceivable way to allow for their increase in numbers was not by a direct creative process but by procreation from a single racial Head, Adam. For procreation, embodiment is essential. And therefore *if man was to be morally free, he must also be redeemable: and to be redeemable, he must be able to multiply by procreation: and to procreate, he must be embodied.*

(3) Now a third consideration enters the picture. It does not appear from Scripture that angels "grow up". They do not start as infant angels and grow up to be adult angels, because they live in a world outside our world of space and therefore presumably outside our world of time. So they do not increase in size nor mature with time as we do.

Angels do not occupy *space* but only *position*. That is to say, no two angels can overlap as it were, by occupying the same position and so confuse their identity. Moreover,

* The theologians say "fallible", but for most of us the word has a slightly different connotation.

since they do not occupy space, they do not need to cross the intervening space to pass from one position to another. The passage is instant. This is no longer such an unimaginable situation for scientific types because modern Quantum Theory sees certain "particles" which are centres of energy as apparently shifting position in just such a manner. These "particles" do not have dimensions and so do not have to pass between the different positions they occupy. The movement from one level of energy to another (it is difficult to express it in any other way than a movement) does not occupy time, any more than these particles occupy space.

So we have no new-born angels, no infant angels, no adolescent angels, no ageing angels, because by reason of the very timelessness of their existence we have to suppose that they are created already mature. It takes *time* to mature, and the maturing process seems to depend on an ageing process. There can be therefore no such thing as the development of character "on the job" – though for the fallen angels there is obviously a sudden destruction of character.

Clearly it has been God's intention that man's role as governor of the Universe is *to engender a maturing process as a by-product*. It is our interaction with the physical world and with one another (physical beings, all) that is to result in "the perfecting of the saints". For this interaction we must have bodies that mature even as we have spirits that mature.

Because interaction with the world is so essential for the process of maturing, it would appear that even the Christian is called upon to react with the world and will not be taken out of the world (John 17:15) until he or she has matured. For this reason it is important that we do not forget to assemble with other Christians not merely in spirit but in body (Heb.10:25). It is not good to go it alone (Gen.2:18) though circumstances sometimes dictate it.

In short, "we are in our relations" both with other things and with other people and these relations need to be physical as well as spiritual. We speak of making *contact* with people without realizing that we are tacitly acknowledging the importance of embodiment. Without embodiment we apparently could not be "made perfect" as God has planned, even as the Lord Himself was "made perfect" in his incarnation (Heb.5:9) where "perfect" means *mature*.

In total isolation from human contact, such as happens now and then with feral children deserted by their parents in infancy and adopted by animals, there is no advancement into humanness if the isolation continues for long enough. All strictly feral children remain non-human in personal development until they begin to interact *with other human beings.*

In a similar way, fully humanized individuals deprived of all sensory (i.e., bodily) input from the physical world can so disintegrate as persons that it proves to be one of the most damaging modes of torture – so damaging that virtually all nations have outlawed it, though unfortunately the agreement is frequently honoured only in the breach of it.

We thus conclude that *embodiment is essential for the perfecting of the saints.*

(4) Then there is a fourth consideration. Granted that embodiment is a practical necessity, we must also have consciousness of our physical environment through the senses – seeing, hearing, tasting, smelling, feeling, and of course all locomotive and manipulative faculties. These are mediated from the will to the hand, foot, tongue, and so forth *via the brain*. It is clear from neurophysiology that the brain is the physical link between the will and the world.

Indeed, there is no unequivocal evidence either from science or Scripture that man can retain consciousness without an actual brain, either in this world or in the next. Angels can, for they were so fashioned, but man was not so designed. The most cogent argument for this must surely be the tremendous emphasis in Scripture on the promise that we shall be "reclothed" in a new *body* (2 Cor.5:1-5). What need of a body if we can function and be perfectly effective persons *without* the body? And what need for the Lord's resurrected body if He, too, can remain truly MAN without it?

True, it will operate on a different principle (as we shall see later), but it will be an embodied existence that we shall enjoy. Just as the Lord deliberately set out to demonstrate to his disciples after his resurrection that He had a body of flesh and bone, so shall we have a body of flesh and bone. He was not a ghost; nor shall we be, for we shall be like Him (1 John 3:2) and we shall have a body like his glorious body (Phil.3:20, 21). *This body will be our conscious link with the new heavens and the new earth; and the source of our "personal awareness" just as it is the source of our personal awareness in this present universe.*

(5) And then there is a fifth reason relating to our personal identity, the matter of *recognition*. We have no idea how angels are recognizable to one another. How does one identify one angel from another when meeting one on a street in the new Jerusalem except they have some formal individuality of shape or size or visible mannerism?

We establish our identity in a dozen ways – by facial features, body shape, size, mannerism, walk, voice, colour: one could probably think of many other characteristics. *The common factor is embodiment!* All such means of identity are bodily; even voice, for we have no voice except we have vocal chords, lungs, throat, tongue – even teeth! Nor could the other party be seen or heard or felt but for the fact that we too have the means to feel and hear and see them, and they us.

The resurrected Lord used his own body, transfigured though it was, to establish his personal identity. It is difficult to think that we who have matured in a physical body and will be punished or rewarded for the "deeds done in the body", could suddenly appear without any shape or form that would enable us to identify one another. I think it safe to say that *embodiment is essential for the preservation of personal identity.*

We have considered five of the reasons why it was necessary for God's purposes that *we* should have bodies. It remains now to consider very briefly three reasons why it was necessary both for us and for God that *He* also should assume a body.

God Himself must become incarnate first, and most obviously, because man needed a Redeemer. A Redeemer must experience physical death in man's place. And such a Redeemer must be of infinitely greater value than any one individual, if his sacrifice is to be sufficient for many individuals needing salvation. No one man, how ever perfect, can be of sufficient worth to redeem more than one sinner.

On the simple principle of an eye for an eye, a tooth for a tooth, and a life for a life, the cost of sacrifice for the millions who need redemption had to be, accordingly, far in excess of the sacrifice of one single human life. Only the sacrifice of God Himself could suffice.

But as Luther put the matter very simply, "God cannot die". As a pure unembodied spirit God could not experience the death of man except by becoming Himself "embodied man". Thus it was not sufficient that God should simply create a perfect man and then send him as our saviour, because such a one, though he might sacrifice himself for the sins of one individual, could not by this means pay the penalty of the sins of countless millions.

Therefore God Himself came as an embodied Man in the Person of Jesus Christ who is the express image of the Father's Person (Heb.1:3), supernaturally conceived but naturally born of a woman and therefore in the likeness of man. The very fullness of God thus dwelt among us bodily

(Col.2:9) in his Person, in order that He might bear our sins in his own human body on the cross (1 Peter 2:24) and as Man, "taste" our death. Such was *his* worth that the death He tasted was sufficient for every man (Heb.2:9).

This was one basic reason why God for man's sake was embodied as Man: that God in Christ might experience death for man. Hence it is proper that Scripture should speak of God laying down his life for us (1 John 3:16), thus purchasing the Church "with his own blood" (Acts 20:28).

But there was another reason. God wished to reveal Himself to man. And how better could He achieve this than by embodiment in the likeness of men to share the vulnerabilities of our humanity – hunger, thirst, fatigue, wounds, and the whole gamut of human emotions save those arising from our fallen state.

And so He came among us, a Man among men, and after three years had so revealed Himself, personally, intelligibly, intimately, that He could say to Philip who asked Him to show them the Father, "Have I been with you so long a time, Philip, and yet thou hast not known Me? *Who hath seen Me hath SEEN the Father*" (John 14:9). As Leo the Great, Bishop of Rome, put it (in 449), "The invisible became visible".

And, finally, a very important reason lies in the fact that embodiment subjects man to stresses, fears, hurts, and limitations that entail temptations quite unknown to purely spiritual beings like angels. Nor can even God Himself have experienced these things. *How, then, could He fairly be a judge of men's actions,* not knowing first-hand the nature of man's temptations?

For this reason, the Father has committed to the Son all judgment "because He is the *Son of Man*" (John 5:27). Had God, in the Person of Christ, not shared the human experience, He could not have acted in complete fairness in judging man's sin because the meaning of our temptations would be experientially quite unknown to Him. But in Christ they were known to the full.

As a supreme example, consider one instance referred to in Matthew 27:34 and Mark 15:23. It was customary for a drink of vinegar and myrrh to be offered to men condemned to be crucified if they were felt worthy of this mercy. The drink was a palliative, and according to Alfred Edersheim it was prepared by a kind of Ladies Society in Jerusalem. It was usually offered to the victim just before the actual elevation on the cross.

Apparently it had been found to bring significant relief against the first terrible pain and shock of crucifixion, and many must have thanked the women for their mercy as the body was wracked by the agony it entailed.

The Lord Himself must certainly have been aware of this merciful provision, but even so as a man He had first to taste it to know for sure what was being offered to Him. And He certainly would know what it was by tasting it, since apparently it was bitter to the tongue. Even in the agony of those moments, having identified its nature He resisted the temptation to find relief and refused it (Matt.27:34). It is true that He accepted vinegar which was later offered to Him (John 19:30), but vinegar was not a palliative and it would almost certainly be needed to enable Him to speak and say the things He had yet to say from the cross. The fluid loss from his body resulting from all the scourging wounds He had received would have placed Him in a state of terrible dehydration and probably well-nigh speechless. The temptation to take the previous palliative must have been almost overwhelming.

It is therefore perfectly in accord with divine justice that the judgment of man should have been placed entirely in the hands of One who was both the Son of God by eternal generation, and the Son of Man by embodiment in time, and who was tempted in his human condition in a way quite impossible for Him in his divine nature.

Thus we have reviewed briefly five reasons why MAN is both blessed and burdened with embodiment. (1) To be an effective manager of the physical world; (2) to enable him to increase in numbers as he takes over his responsibilities increasingly; (3) to make the resulting experience of *physical interaction* the basis of a maturing process for his *spiritual perfecting*; (4) that we might have a full consciousness of our world and self-consciousness of ourselves; and (5) that we might have a means of retaining our *personal* identity.

And we have looked briefly at three reasons why God Himself assumed embodiment. (1) that He might by the sacrifice of Himself secure man's redemption; (2) that He might reveal Himself to man in all the beauty of his Person; and (3) that He might be man's Judge with perfect fairness when the day of reckoning comes.

We shall in the chapters which follow have occasion to return to some of these points in more detail. My object will be to underscore in every way possible the fact that man is not embodied by accident but by design, and to show that his body is as essential an element of his very being as his spirit is.

It is literally a fact that having no body is tantamount to being a NOBODY, and that apart from the existence of these millions of 'somebodies' the Universe is nothing more than a gigantic but pointless display of wasted and wasting energy. It is the embodiment of man that gives meaning to the present Universe and will give meaning to the New Universe which is yet to succeed it.

Chapter 3

THE HUMAN BODY

Designed as an Instrument for the Human Spirit

To any thoughtful observer the human body, even in its present state of imperfection because of abuse and disease, must still appear to be the most wonderful piece of machinery in the animal world. It is an instrument uniquely designed to give expression to the human *spirit* in all its moods. And even the evolutionists would admit that between this human spirit and animal spirit there lies a seemingly unbridgeable gulf. The evidence of this is overwhelming to the open mind.

One has only to watch at close range the hands of a piano virtuoso playing a composition by Tschaikovsky, with the fingers striking as many as twenty keys per second across a keyboard of 88 alternatives, to appreciate something of the manipulative skills in the human body.

Consider not only the creation of the music to begin with as an act of the spirit, but the superb engineering of the grand piano with all its technical refinements and artistic embellishments. Then add to this the development of the means of telecasting the performance in colour and movement, providing close-ups of those fingers so clear as to reveal the very texture of the skin, and projecting this image over thousands of miles – to recognize what the combination of head and heart and hands in man can accomplish.

Consider the performance itself. The eye of the pianist rapidly scans the score, seemingly without reference to the keyboard, while his ear monitors the touch and the timing, and his brain interprets the symbols on the page and directs both hands unerringly to the proper positions. The total performance – original creation, provision of means for reproduction, transmutation of the sound waves into radio waves, and the sending and receiving of these waves and their faithful reconversion into the original sounds – all these achievements are entirely dependent upon the interaction between a human spirit and a human body within a physical world. No link in this chain can be omitted.

Even the invention of musical scoring and the very tuning of the instrument itself are involved in this performance. Each requires perfect coordination. Put together, this is an achievement which demonstrates the truly extraordinary capabilities of the human spirit and the human body in producing an astonishing total performance. The number of messages that are flashing back and forth within the nervous system at the speed of light, in both performer and listener, must be reckoned in the billions: and yet the whole system can actually be expected to work time after time almost flawlessly.

Head, heart, and hand are involved in a total coordination that all too often we accept without amazement. Why? Because it is so dependable! Man has not yet produced a machine which even approaches such capabilities. This total artistic and technical achievement would be utterly impossible for a mere angelic being – and, dare I say it, even for God Himself, *unless incarnated*. Would it be altogether absurd to add, "And although God can sing (Zeph.3:17), yet He could not write the score without human hands". It was a finger that wrote the Ten Commandments and a hand that wrote on Belshazzar's wall . . . The words of George Eliot are *á propos* in this connection. In her poem *Stradivarius*, she wrote:

> 'Tis God gives skill,
> But not without man's hands:
> He could not make
> Antonio Stradivari's violins
> Without Antonio.

Perhaps it would be more correct to say, "He will not make . . . ." rather than "He could not make . . . ." for it is by God's choice that He has decided to leave such things to us – not because of any limiting necessity imposed upon his omnicompetence but perhaps because He desires our company.

The whole performance – artistic, gymnastic, and techical – is so extraordinary when one stops to think about it, as to be little short of miraculous. The whole of man is totally absorbed in such an achievement. Without the body to support the mind, and the mind via the brain to direct the body, none of this could be possible. And it would surely be patently absurd to say, 'Oh, an animal body could probably come close enough if properly trained'. The human body is no more an animal body than a human spirit is an animal spirit. The two are permanently wedded teams operating at entirely different levels.

It is obvious that the genius of the composer would not be made apparent without the player and his piano! Nor the skill of the pianist and the perfection of his instrument without the creative genius of the composer. Such accomplishments are interdependent; as they are in the design and erection of a Gothic cathedral, or – to move into another area – putting a man on the moon and bringing him back again with mathematical precision. Man's creative spirit and manipulative skill combine to produce near miracles, and unlike animals man has a delightful consciousness of the achievement.

In an almost infinite variety of forms the potential of the human spirit and of the human body are found matched in every imaginable permutation and combination. When Christians speak easily of the re-birth of the spirit without also telling of the redemption of the body, they are speaking only of the saving of half the man – which is really no salvation at all, for MAN.

This one example of the capacities of human beings could be multiplied almost *ad infinitum*. It is not that man can fly with the ease and manoeuverability and precision of a bird, or run like the cheetah which has been clocked at 70 miles per hour, or swim with the speed and grace of the dolphin, or jump like the kangaroo or gazelle, or perform the aerial acrobatics of the monkey or squirrel, or scale the mountain cliffs like the goat or mountain sheep. In such particular achievements, essential to the survival of these animals, they often easily out-perform the capacities of the human body. But they do not, except on rare occasions, even in these achievements out-perform what man *can* do by his combination of inventive spirit and unique body.

The capabilities of man are almost unlimited and they are freed from the necessities of mere survival. Indeed, such is the spirit in man that he is even willing to elaborate his culture to the point where it cannot survive! If we exclude such aberrations, the embellishments with which man beautifies or seeks to beautify his world are a reflection of the manifest delight which God Himself took in exhibiting his own creativeness. They demonstrate a kind of common grace that smooths the troubled path man must now follow because of his fallen nature. Without his body man would almost certainly be not one whit more creative than angels appear to be. Only God and man are creative in this sense: God because He is God, and man because he is MAN, a *human* spirit in a *human* body – both of them designed and created by God.

With the creativity of his mind and *its* brain, the acuity of his stereoscopic vision, his refined manual dexterity, his easily maintained erect posture, his vastly more versatile nervous system that makes his body an extension of his mind, he can achieve all kinds of exceedingly complex tasks which often, by reason of his skill, appear quite simple. Actually, they are far beyond the capabilities of animals – such as threading a needle which even the most "educated" chimpanzee cannot do!

Now the idea that the *spirit* in man is a direct creation of God is very ancient and strongly supported by Scripture. Almost all theologians, Protestant and Roman Catholic alike, agree to this general thesis in Adam's case, at the very least. But a very large majority would go one step further and say that each individual spirit is still being created, one by one, and infused into each individual human body at some early stage in its development in the womb or at the very latest at the time of the drawing of the first breath.

Thomas Aquinas (1226-1274), one of the giant intellects of Christian Medieval times, argued that the soul in each case is specifically *designed* by the Creator to suit the particular body for which it is intended. Body and spirit are thus matched, not merely in a general sense but in a particular sense in each case. And a number of modern theologians, both Protestant and Catholic, support this thesis and find it, too, clearly reflected in Scripture.

If God is sovereign and has appointed to each of his redeemed children a specific life work, and if each of us is a duality of body and spirit, then it follows of necessity that both the genetic endowment of the body and the life experiences that *mold* the spirit, must equally have been divinely ordained – as must also the *nature* of the spirit which God creates.

Task and talent have to match if the plan is to work out. As A. H. Strong in his *Systematic Theology* put it (quoting W. Gladden): "Heredity is God working in us, environment is God working around us". God never calls us individually to a life work for which He has not also equipped us both physically and spiritually.

In a very ancient book, *The Testament of Naphtali*, we find this observation (2:2-4): "As the potter knoweth the vessel, how much it is to contain, and bringeth the clay accordingly, so also doth the Lord make the body in accordance with the spirit and according to the capacity of the body doth He implant the spirit . . . . and as the potter knoweth the use of the vessel, what it is meet to be used for, so also doth the Lord know how far it is capable".

This extraordinary combination of body and spirit that is man, is in fact so extraordinary as to constitute a new "thing" in the animal world. He appears suddenly on the scene, not just as a continuation or routine link in the great chain of being with refinements that are merely *quantitative*. These refinements are *qualitative* to such an extent as to constitute a virtual discontinuity in any imagined evolutionary process. The geneticist Richard Goldschmidt proposed that a sudden jump of this kind should be called a *saltation* since it clearly involves much more than a mutation. George G. Simpson felt that this term came too near to the supernatural concept of creation. So he proposed instead the use of the term a *quantum leap*, borrowing from physics which thus gave it a more respectable parentage!

Today even this term is unfashionable and has been replaced by the current phrase *punctuated equilibrium*, brought into popular favour by Stephen J. Gould. It means simply that the normal course of evolution proceeds by very small formal shifts that scarcely rock the boat until suddenly a dramatic discontinuity occurs to "punctuate" the smooth course of events. But a rose by any other name will smell as sweet, and that is all these terms are – roses by other names. It would be difficult to distinguish an evolved species from a created one in the fossil record!

Each of these new phrases is manifestly the old one spelled differently . . . . They tend to be presented to the public in the guise of new explanations, whereas in point of fact they have no explanatory value whatever. They are old descriptions, not new explanations. In no way do they *account for* the sudden appearance of man in all his tragic glory. They merely demonstrate that man's coming established a genuine discontinuity.

Professor Suzanne Langer who is no friend of the Christian viewpoint, speaking of language as one of man's singular achievements, put the matter thus: "Language is without doubt one of the most momentous and at the same time the most mysterious product of the human mind. Between the clearest animal call of love or warning or anger, and a man's least trivial word, *there lies a whole day of creation*" (emphasis mine).

In a somewhat similar vein Humphrey Johnson wrote: "There is a wider difference between a man and a gorilla than there is between a gorilla and a daisy". Such statements could be multiplied from many sources. J. Fiske, an early contender for the evolution of man's body, as quoted some years ago by James Orr, remarked, "While for *zoological* man you can hardly erect a distinct family . . . . for *psychological* man you must erect a distinct *kingdom*, nay, you must dichotomize the universe – putting man on one side and all else on the other".

It is true. Man stands apart from the rest of nature. And contrary to Fiske's admission, his apartness relates to his body as well as to his pysche, since without this body such a spirit would be impotent, while such a body without such a spirit could only be a total anachronism in the evolutionary scale of things.

The concept of a spirit that is specially suited to a body is an ancient one which we shall look at later, but it is worth noting at this point because we all too often assume that the only thing God is concerned with perfecting in his people is their *spirit*. This presupposes that the spirit can stand by itself and will come into the presence of God by itself. But as we shall see increasingly in the chapters which follow, this is not at all the case.

It is obvious that Adam's body was created before his spirit was created. And probably this is the case in every individual. But this does not signify that the spirit is then poured into a vessel whose shape has arisen by chance. Both the vessel and its content are designed with a single purpose in view for the individual. Certainly a body can exist without a spirit, as it did in Adam's case while his body lay on the earth awaiting the breath of God, and so it may do in ours – for a few hours after our spirit has fled, for example. This potential for independence of each component from the other leads me to suggest an analogy regarding the human body as a vehicle designed ahead of time for the human spirit which is to animate it.

If a man builds a house for his animals, he suits its construction to their nature, besides being guided by what he hopes to do with them. If he was raising snakes for their venom, he would build a house they could not escape from; for his cattle he obviously builds a much larger house from which they can readily be allowed out; for his horses, the egress must be more carefully managed since they are vagrant creatures by nature. For his dog he would construct a house that would in some measure share his own home comforts since this is what the dog will probably do during much of its life.

Thus the nearer he gets to a house for a creature sharing his *own* nature, the more nearly will its total accoutrements resemble his own house. And as to his own house, how does he build it? As far as he has the means, he will build it to suit his own nature. To some greater or lesser extent he will seek to satisfy the natural inclinations of his wife and his family, but fundamentally if it lies in his power to do so, the builder will build it as a vehicle for the expression of his own person.

Now what, then, will God do if *He* decides to build a house which is to be fit for Himself, which in due course will be *his* habitation, a house which is to serve Himself for thirty-three years, in which *He* will live and express his character, occupying it day and night, constantly, actively, fully, sleeping and waking, being born and dying? It will be a house capable of being so lived in, appropriately and worthily. It will be house that can sustain the demands of habitability that *He* will make upon it.

It will be beautiful because God clearly loves beauty, having created so many beautiful things in nature. It is difficult to see how the beauty of many creatures can possibly serve any mere survival purpose, while many very ugly creatures (especially insects) survive and multiply very freely! Moreover, it must be flexible enough to allow the whole spectrum of human mood from delight to near desperation, from a groaning within to a sudden exclamation of glad surprise, for it must make communication by gesture or tone of voice, or even "turning to look" in sorrow and reproof, or "turning in anger". For the body is by no means without its own powers of communication. And it must be kingly enough that worship at the proper time is both naturally accorded and accepted with dignity.

Finally, and even more importantly, it must be such a house that while it will never of itself wear out, it can nevertheless be deliberately sacrificed when the proper time arrives.

All of this, of course, points to the Incarnation. It was just such demands that were to be thrust upon the body of the Last Adam for which preparation was made in every particular by the creation of the body of the First Adam. And these capacities must therefore apply to the body of the very first human being as they must to the very last human being – as we have already noted. If this is not so in the most complete sense imaginable, then the Last Adam surrenders his right to that title and can no longer stand as substitue for the First Adam to act as the new Federal Head of the redeemed family of man.

Aristotle wrote, "The nature of man is not what he is born *as* but what he is born *for*". If I may convert this into

Christian terms, it could be re-written as: "the body of man is not what he is born as *now*, but what his body was designed for *then*".

Tertullian has a wonderfully descriptive passage in which he depicts the Creator bending over his clay as He eagerly fashions man's body. "The truth is, a greater matter was in progress, out of which the creature under consideration was being fashioned. So often, then, does it receive honour, as often as it feels the hands of God, when it is touched by them, pulled by them, drawn out, and molded into shape. Imagine God wholly employed and absorbed in it – with his hand, his eye, his labour, his purpose, his wisdom, his providence, and above all, his love which was dictating the lineaments of this creature".

Tertullian concluded, "Whatever was the form and expression which was then given to the clay by the Creator, *Christ* was in his thoughts as one day to become Man, because the Word, too, was to be both clay and flesh, even as the clay (in the Creator's hands). For so did the Father previously say to the Son, 'Let US make man in OUR image, after OUR likeness'. So God made man, that is to say the creature which He was molding and fashioning, after the image of God – or in other words, after the image of *Christ* did God make him . . . . That clay which was even at that moment putting on the image of Christ who was to come in the flesh, was not only a work of God but actually the pledge and surety of God (for the redemption of man)".

Such a house for the spirit of man, like Solomon's Temple, was not merely to be like any other pagan temple already in existence, any more than Adam's body was merely a copy of some other animal body already in existence. It was to be exceptional, "exceedingly magnifical" (1 Chron.22:5) as the King James Version quaintly puts it!

And originally it must have been glorious indeed. Imagine a human body which, despite the defilement of sin to which it was to become subjected all too quickly, nevertheless survived with all its energies largely unimpaired for nearly a thousand years! The body in which Jesus Christ took up residence for some thirty-three years was Adam's original body recovered – and it, too, was "magnifical".

The divine Architect had designed it for Himself in the first place, so we may be sure that the body of the Lord Jesus Christ was not the tumble-down house in which we struggle through life. His body magnificently supported Him daily as He lived out his life among men; and it provided perfectly all the resources for the expression of his divine nature. His presence in the body was so magnificent that even the most callous of his enemies had to step back sometimes in awe, and they only had the courage to abuse Him because He deliberately veiled his glory, and allowed them to do so.

Undefiled by sin and indwelt by the Lord Himself, a superb human body appeared on the stage of human history and men worshipped without shame or hesitation the One who possessed it. I suspect that in our present sinful state we might easily have fallen down and worshipped Adam as he came from the hand of God – such was the glory of *his* body.

Evolution can present us with nothing comparable which could serve as a prototype for the Last Adam.

Chapter 4

THE IMMORTALITY OF A HUMAN BODY

Biologically and Theologically Considered

When Adam and Eve ate the forbidden fruit they did not merely shorten their lives and so die *prematurely*. They introduced into their bodies an entirely new and foreign element – MORTALITY.

Does this mean that if they had not eaten this fruit they could have lived on FOR EVER, in the same body? The answer from Genesis 3:22 is, *Yes!* But this affirmative answer needs qualification. Because, while they COULD have thus lived on for ever, we learn from passages of Scripture elsewhere in the Bible that these bodies of ours have yet a further state of being awaiting them, a state which involves a transmutation, a kind of "graduation", to a still higher level of wholly indestructible bodily existence. There are thus two levels of immortality: one which signifies that the body need not die though it *can* be deliberately put to death; and one which signifies that the body cannot die under any circumstances whatever.

Genesis 3:22-24 reads as follows:

> And the Lord God said, Behold, the man is become as one of us, to know good and evil: and now

> lest he put forth his hand, and take also of the tree of life, and eat, and live for ever . . . .
>
> Therefore the Lord God sent him forth from the garden of Eden, to till the ground from whence he was taken. So he drove out the man; and he placed at the east of the garden of Eden . . . . a flaming sword which turned every way to keep the way of the tree of life.

Now this tells us that although the forbidden fruit had already done its fatal damage in the bodies of Adam and Eve, this damage could still have been undone so long as they remained in the Garden and had access to the Tree of Life. For by eating from the Tree of Life they could evidently have been healed of their acquired mortality and gone on living for ever. By which means their *bodies* would have been healed – *but not their corrupted spirits*. The Tree of Life reappears in Revelation 22:2 and we are informed that its leaves were for healing . . . .

The precaution cannot have been to prevent recovery of *spiritual* life from the Tree of Life (as some have proposed), for surely this would not have been any greater danger than we are in at the present moment with our spirits regenerated even while our body yet remains to be redeemed. It must therefore have been the danger of *physical* healing without spiritual healing. This was what created the emergency.

That it was an emergency seems clear both from the fact that Genesis 3:22 is one of the very few unfinished sentences in Scripture – as though the divine Author caught his breath at the very thought of the consequences; and from the fact that expulsion is emphasized in verses 23 and 24 by the words *sent forth* and *drove out*. Thus two physically immortal creatures had now become mortals, and their very mortality was a measure of protection against the effects of their fallen nature, for death was now not merely a penalty but a remedy. It rid them of their "body of sin", as it will deliver us also.

But one may ask, In the light of modern biological knowledge, is it really likely that a physical body of *any* kind could be so constructed that it would live on and on and on for ever, never to wear out and break down? And, surprisingly, the answer is, Yes, it is perfectly possible!

Let us look more closely at this almost incredible fact and set forth the evidence for the assertion that a physical body really can have the inherent potential for unending life – barring accidents.

Since the subject has been discussed at considerable length in another volume by the author,* it will only be treated briefly here, sufficiently to enable the reader to perceive the nature of the evidence and how it is to be applied in the present case.

It is most important to recognize that physical immortality does not mean that a body with such a constitution cannot die. An immortal creature CAN be killed. The basic distinction between a mortal creature and an immortal creature is that a mortal creature will die in due time as a matter of course, whereas the immortal creature *can* die by being killed but NEED NOT die if certain conditions of life are maintained.

Of course, you may say, "Well, that's not what I understand by the word immortality". And the remark is quite justified since immortality is commonly used to mean deathlessness of the spirit rather than the body. But for the biologist, physical immortality means only that death is not inevitable, not the natural consequence of being alive, not the *destiny* of a living organism, but something purely external to its possession of life.

In biblical parlance there is a bodily immortality of a lower order, a contingent immortality dependent upon certain conditions, which exists as a prior stage to a bodily

---

*** See *Seed of the Woman*, Doorway Publications, Brockville, Ontario, 1980, especially pages 3-29, and 75-97.**

immortality of a higher order that is absolute and beyond commutation. The higher order of immortality belongs not to biolgoy but to *theology*.

It will be helpful to illustrate this by specific reference to living organisms which exemplify the lower order of physical immortality and then to show how the higher order of absolute immortality differs from this lower order.

There are billions of living creatures in the world today which are strictly immortal in the biological sense. These creatures never die a *natural* death as far as is known. When they grow to a certain size (approximately twice their beginning size), they simply split in two and go merrily on their way as two individuals – never experiencing death nor leaving any corpse behind. They are their own parents: the "father" does not give birth to a son but becomes his own two sons. The process involves no *birth*, only a simple division of one living organism that divides and becomes two; and the process goes on unendingly. These creatures are unicellular and well known to biologists as amoeba and paramecium, and some other forms.

Since each one divides into two and then grows to double size and splits again, they could soon overwhelm the earth were it not for accidents that happen to them, chiefly physical injury or predation. But they do not die for inherent reasons and are considered by biologists to be strictly immortal on that account. There are billions of them in the world, creatures who live without any prospect of dying barring a fatal accident.

Now it may be objected, "Well, that's all very well. But they are little blobs of life and exceedingly simple unicellular creatures . . . . That's very different from a higher animal like a man, for instance". Quite true: yet neither the amoeba nor the paramecium is as simple an animal as it would appear.

These small creatures have been studied for over a century. One of the most famous investigators of their behaviour was H. S. Jennings who in 1910 published a book on his findings. He had been observing them for some years, spending hours on end with his eye glued to a microscope. His conclusions are remarkable.

He discovered that amoebae displayed signs of highly advanced forms of behaviour which, as he put it, were they to be magnified to the size of a dog they could only be interpreted as *anger, determination, frustation, hesitation, attentiveness,* and according to Jennings, even *intelligence!*

Another much more recent observer, J. Boyd Best, fully confirmed Jennings' conclusions, and added to the list such emotions as *boredom, rebellion,* and even "*cognitive awareness*" (which effectively is simply consciousness)!

Such immortal creatures do indeed display a remarkable measure of individuality which could almost be termed personal identity, and yet they go on living indefinitely. And remember that we are speaking of living forms of animal life which are, despite their unicellular nature and microscopic size, very much alive. We should not be deceived by their size – for as Nicolas Malebranche remarked when he looked through a microscope for the first time, "This is the end of size"! After all, how *big* is life?

Thus we can easily establish a point that comes as a surprise to many people: a creature of such sensitivity can still be immortal . . . . It demonstrates unequivocally that physical immortality is "a fact of life". As Professor H. J. Muller put it, very simply, "Natural death is not the expression of an inherent principle of protoplasm". And Julian Huxley reaffirmed this when he wrote, "Functioning protoplasm is not in itself mortal".

By protoplasm is simply meant the stuff of life, and when it is said that protoplasm appears in forms that are effectively immortal, it does not means that such forms *cannot* die, it only means that such forms *need not* die. Protected from mortal hazards external to them, these creatures simply *do not* die. They just go on dividing and multiplying *ad infinitum*. It is only accidental death that prevents them from overwhelming the earth. They do not die of old age – as we do and as most animals familiar to us do as a matter of course, including our pets. They never die a "natural" death.

Now in the case of Adam, God evidently endowed his body with just such a property as this, a potential for endless continuance. The processes of self renewal and repair could have gone on for ever. It is clear that he could die, since we know that he did die – though not without first surviving for almost a thousand years. But it is equally clear that he would never have died if he had not sinned. If this were not true, the penalty threatened for disobedience would have been no threat at all. One cannot discourage disobedience with a threat of "punishment" in a form that will happen anyway whether there is disobedience or not.

It has been suggested that perhaps the threat was not death per se but rather *premature* death. But this interpretation is entirely unsatisfactory because it would imply that the death of the Last Adam was also premature *and nothing else*. Any life given prematurely is not life given vicariously but merely a life shortened. But we know that the Last Adam was, like the First Adam, made *after the potential* (Greek: *dunamis**) *of an endless life* (Heb.7:16), and this forces us to conclude that the First Adam must also have been in this position. It was not premature death that Adam introduced; it was death itself. "By one man sin entered into the world and by sin death . . . . (Rom.5.12), and this acquired mortality became the lot of all of Adam's descendants naturally born.

Adam and Eve had to be created with bodies capable of

---

* So Professor Walter Grundmann of Dresden, who suggests the words 'possibility', 'capacity', 'ability', i.e., according to the potential of (in *Theological Dictionary of the New Testament*, Vol.2, p.285).

endless continuance and under no necessity of dying in order that the Redeemer of man's body might Himself likewise be under no necessity of dying, while yet remaining truly representative of man as created.

In the strictest sense, therefore, we must re-define immortality as it applies to the body of the First Adam and the body of the Last Adam. It was *contingent* immortality, an immortality dependent upon the fulfillment of certain conditions, but a real immortality in the true biological sense. And the contingent condition for its realization was lost in the Fall, but restored in Christ by the Virgin Birth.

While the contingency hinged upon obedience, it also hinged upon the fact that our bodies, like those of the humble amoebae, are *vulnerable*. We can be fatally injured. We are not impervious to mortal wounds. The processes of life which in Adam could have continued to function for ever were nevertheless open to irreparable damage.

Both Adam and Eve could, and did, die in due course: but had they been obedient, such a terminal event need never have occurred. Their death was not an *accident* in the sense that it is for the amoebae, it was a *penalty* imposed for deliberate disobedience, the penalty being a newly introduced internalized defect in their constitution. They died for inherent reasons, from damage to the machinery of life resulting from ingesting a fruit with fatal consequences inherited by all their descendants, save One.

So much, then, for the lower level of immortality which we have referred to as contingent immortality, which is both a biological and a biblical concept. But there is another and higher form, an immortality to be achieved when we are raised from the dead in a more glorious and wholly in-vulnerable body. This higher form of immortality is *absolute*: we shall be placed in the position of neither *needing* to die nor even being *able* to die! Death will be a thing of the past. This is no longer, of course, a biological concept but a theological one: yet it applies strictly to a resurrected *body*, a body as real as the resurrected body of the Lord Jesus Christ.

Augustine (340-430) thought deeply about these things. In his characteristic way and by the effective use of aphorism he wrote of the nature of Adam's body as created:

> It was not impossible for Adam to die
> but it was possible for him not to die.

As he later reflected further upon the matter, he realized that there is an even higher form of immortality awaiting man, at which time we shall find that:

> It is not only possible *not* to die
> but it will be *impossible to die at all*!

Such is the goal for all the Lord's people, for "He that wrought us for this very thing is God Himself" (2 Cor.5:5).

In summary, then, Adam's body was endowed with a potential for unending continuance. As the biologist would say, his life was not "spanned", no limits were placed upon it provided that he obeyed certain injunctions.

This potential or contingent immortality was necessary in order that in the event of disobedience he could still be redeemed by provision of a Saviour who would be in a position to taste of death in his place. For this purpose, the Redeemer's body must (1) be truly representative of Adam's body and therefore truly human, and at the same time (2) under no necessity of dying for any inherent reason.

The latter was as important a requirement as the former. The Saviour must fulfill both conditions. The first, in order to be truly representative of Man as God intended him, and the second in order to be able to offer himself by a death that was vicarious and not merely premature. This situation predetermined the nature of the First Adam's body.

Evolution holds that all higher organisms have built-in

spanned lives and are *programmed* to die after a certain length of time in some way determined by the nature of the species. This is necessary, they hold, in order to allow evolution to work, since it is essential to keep the way open at all times for the new and more advanced forms to survive in their initial stages of development if they are to compete and replace the older and more established forms. Only so can there be guaranteed the linear progress believed to characterize the evolutionary process. Without death there would be no room for replacements, and without replacement there would be no evolutionary advance.

Evolution cannot account for the emergence of *man*. For man alone possesses a constitution which guarantees his continuance *permanently* in the web of life if certain conditions of obedience are fulfilled.

Now, the death of the Redeemer must be vicarious, not merely premature. Premature death is a death which may, for example, take the form of a soldier's sacrifice at the front, or a suicide, or a youth in a car accident – each dying "before the time expected". Vicarious death in the theological sense never has any element of dying "before the time expected". Death is *never* "expected" in the sense of being normal, for One who is to die vicariously.

This is true of vicarious death because it can only be applied to the death of an immortal creature who has no expected time of dying. Premature death presupposes something inevitable in due course, and therefore applies only to a mortal creature. Such a death is a mere "cutting short" in the midst of life. Substitutionary death is not a cutting short but a "cutting off" (Dan.9:26), the sudden cessation of a life which could have continued for ever but was deliberately terminated for the sake of someone else.

Since death was in no sense inevitable for the Last Adam, it cannot have been inevitable for the First Adam. Otherwise the Last Adam was not truly representative of the First Adam, i.e., of MAN as God created him.

To be executable was essential, in order that He might die substitutionally: but this executability was related to what could be imposed from outside, not what was programmed from inside. A vicarious death is never the same thing as a premature death. These are entirely different categories of experience.

Chapter 5

A WOMAN IS BORN OF A MAN

**And the Lord God caused a deep sleep to fall upon Adam,
and he slept;
and He took one of his ribs and closed up the flesh thereof;
and the rib, which the Lord God had taken from the man, made He a woman
and brought her unto Adam.**
Gen.2:21,22

From the Christian point of view, the formation of Eve out of Adam was a *biological* fact of tremendous *theological* importance. From an evolutionary point of view, it is a sheer impossibility and nothing more than a piece of imaginative nonsense. From the point of view of the Creator Himself, there may well have been no alternative.

It may surprise the reader that had Eve been a separate creation and *not* formed out of Adam, she could not have shared in Adam's redemption! Nor could *her descendants!* It is necessary to emphasize "her descendants" because her title as the "mother of all living" (Gen.3:20) is just as crucial to the Plan of Redemption as Adam's title "the father of all dying" (1 Cor.15:22). The reasons for this will become clear later.

With this as a kind of summary statement, let me set forth as simply as possible, without departing from well-established fact, what I believe to have been the circumstances from both a theological and a biological point of view. The record forms a meeting place of profound importance between revealed truth and scientific fact.

Each statement in the following discussion is capable of documentation but it was felt desirable throughout this small volume not to encumber the text with footnotes and call numbers, but to depend on the reader's confidence in the writer's assurance that there are no unsupported assertions made where physiological data are introduced.*

When Adam was first created and introduced into the world, there may very well have been a number of man-like creatures already in existence, the result of creative activity before the fashioning of Adam. And it would seem reasonable to assume that some of *these* creatures were among those presented to Adam as potential mates.

That not one of them was truly human is borne out by the fact that not one of them was accepted by Adam. It is a law of nature, clearly established by the Creator to preserve order, that no species will accept a mate from any other species, no matter how similar in appearance they may seem to be. Thus even the most likely candidates by our judgment were thereby proven not to be *human*.

By this means it was now made clear that only those who were, effectively, "in Adam" could be acceptable as mates for a human being, and only these were in the future to be counted as true members of the Adamic family. Whatever covenant was made with Adam as Head of this family of Man, that covenant was thereafter applicable only to members of the Adamic species. This Adamic species was the sole subject of all blessings and cursings which were to follow – except in so far as there were repercussions throughout the whole of nature.

Now these observations apply also to Eve whose relationship to the family of Adam is quite unique. It has to be borne in mind that by definition a species is an interbreed-

---

* Very complete documentation will be found in the author's *Seed of the Woman* published in 1980 and obtainable from Doorway Publications.

ing community which naturally produces fertile offspring. Such a community by general agreement is always viewed as a family of related individuals who are all derived from a single parentage – by evolution according to the evolutionists, and by creation according to the creationists.

As we have already seen in Chapter 4, angels do not multiply by propagation as man does but as a direct result of the creative activity of God. They neither marry nor are given in marriage (Mark 12:25). Thus *as a class of beings* they do not form a single species: instead, each individual becomes a species in itself, and there are no family relationships between them. It follows that they have no ancestral Head, no single representative, no "First Angel" from whom all other angels are descended. Since the only Plan of Salvation of which we have any knowledge involves a Saviour who assumes the position of Headship of a new family, the Bible gives us no clues as to how angels as a class could be redeemed. If such is possible at all, we have no model. Logically, it would seem that a separate saviour would be required for each individual angel, millions of saviours for millions of angels.

Bearing in mind therefore that, because direct creation produces separate species whereas procreation produces families, it is clear that while Adam as a first man had in the very nature of the case to be created, his descendants are subsumed under his Headship by reason of their procreation. All are his relatives *by descent.*

What, then, is to be done with Eve? How can she be created *separately*, like Adam, without being *unrelated* to Adam – and therefore a separate species no matter how alike physically in all appropriate respects? Adam and Eve under these conditions of origin would also be precisely what angels are – unrelated *separate species*. All of their descendants would therefore enter the category of hybrids rather than a pure race.

What, then, becomes of the Headship of Adam over his family? Would we not have, in fact, two Heads? And must there not therefore be two Second Heads, a "second" Adam and a "second" Eve? Remember that we are dealing with a real situation, a critical moment in the history of the race. And one of the best ways of assessing the true importance of such a moment is by logically considering the conseq-quences.

Only those can be redeemed who by reason of being "in Adam" are in the lineage of Adam and counted as his seed. The redeemed are always of this species. No evolutionary antecedents, if human evolution were true, can qualify as redeemable – nor any of those other species who may have been his contemporaries and continued to share his world.

But someone had to be a help-meet, a mate to be his partner in the propagation of the species. If such a partner must belong to the Adamic species, a *created* Eve would not do, for all of Adam's species must be one and "in Adam" to qualify as redeemable. It is in accordance with this fact that Paul said, "God hath made *out of* (so the Greek) *one* * all nations of men . . . . on all the face of the earth" (Acts 17:26). And it should be noted that it says "out of one", not "out of a pair".

Then how were our first parents to be constituted so as to form a pair "out of one" without two separate creations? The secret of our truly human identity lies in our all being "in Adam", whether for good or ill. This must include Eve. *In order that Eve might also be "in Adam" it is clear that she must be taken out of Adam as to her origin: she cannot have been either evolved independently – nor even created independently.*

The taking of part of Adam for the formation of Eve's body, while Adam was in a state of deep sleep, was tantamount to a process of divine surgery under anaesthesia.

---

* The word *blood* almost certainly does not belong in the original text.

The "closing up of the flesh thereof" which left Adam in some way reduced but whole nevertheless, must signify that after the operation his body was significantly different. A radical change in the functioning of his body had now come about.

The divine surgery involved in the formation of Eve may quite possibly have had nothing to do with a "rib" * at all but only with some structure that was on one side of Adam's body, since the word for *rib* could equally well mean a *side member*.

Presented with Eve when he awoke, Adam was apparently immediately conscious of a creature born "of his flesh and of his bone"; a creature somehow once part of his very self who was now "the other half".

It was as though, before her formation, Adam was a whole *person* and a whole *man*; but now that Eve was formed from part of him, though he was still a whole person, he was only *half* a man. This seems to be the sense of the words in Genesis 2:24 that together they shall be "one flesh", since the word for *flesh* in the original Hebrew (*basar*) in the Old Testament never has any other meaning than that of *body*. It does not signify 'lower nature' as it may sometimes in the New Testament.

It may be thought the concept of an Adam somehow combining within himself both male and female, as would seem to be implied, is a repugnant one. But it must be borne in mind that to a greater or lesser extent this is true

* Regarding the "rib": precisely what it was that God took from Adam for the building of Eve has long been a matter of dispute among commentators. In the volume by the present author mentioned above, there is an extended excursus on the identity of the "rib" in the light of ancient traditions and more particularly of Assyrian and Babylonian cuneiform words and ideographs for "woman". Some thought is also given to the possible etymological development of the Hebrew word (*tsela*) rendered "rib" in most versions. Perhaps the simplest explanation of what occurred in this surgical operation is that sexual dimorphism was initiated in man. It is also conceivable that a very similar process accounts for sexual dimorphism wherever it is found in every other animal species, if they, too, were all direct creations.

of all of Adam and Eve's descendants, including ourselves. Yet it must also be borne in mind that wherever there arises a "confusion of gender", it can only be described as an aberration, if not a pathological condition.

As man is now constituted, resulting from the separation of Eve out of Adam, the two sexes have been divinely allotted to two differently constituted bodies – without any such aberration under normal circumstances. Adam as created was a perfect creation. The union of the two principles of maleness and femaleness (both hormonal and functional) must in him have been perfectly ordered. When these organs (and the hormones they generate) were separated and appropriately re-housed, two equally perfect bodies resulted. It is only when, due to some fault in the mechanisms of development within a particular individual an aberrant form emerges, that we are distressed by it. Abnormal re-union of structure, which God has designed to be separated, is bound to be an aberrancy. Such aberrations are not strictly a fusion of male and female such as must have been in Adam, but a *con*-fusion such as must inevitably arise when fusion occurs contrary to what God intended.

It must be borne in mind that in Adam's undivided body (i.e., before the formation of Eve) the two elements compounded in one organism may have produced an internal organization different from anything which exists at the present time – save under very exceptional circumstances. It often happens that combinations of elements produce results quite different from either element alone.

Originally Adam may well have had a form which *did* accomodate maleness and femaleness perfectly. After all, Adam was a creature formed to reflect physically the personal nature of the Creator Himself who is spiritual, in whom there is no division of things which we now view only as antithetical. We know from Scripture that God is presented in the role of both Father and Mother – as Father, for instance, in 1 Chronicles 29:10 and Isaiah 64:8;

and as a Mother in Isaiah 49:15 and 66:13. This applies also to the Son as is evident from Isaiah 9:6 and Matthew 23:37.

Moreover, it is a well established fact that the embryo at first exists for some time *in a sexually undifferentiated condition*, exhibiting the potential for development in either direction. Then hormones begin to take over and drive the organization of the growing body towards one pole or the other. If it should be argued that the X and Y chromosomes have already determined which way things are to go, two qualifying factors must be recognized.

In the first place, every male body carries both X *and* Y chromosomes in every cell in the body and therefore they are available as triggering devices in either direction. And in the second place, these X and Y chromosomes are now known *not* to be absolute determinants: there are other directive agencies at work which do on occasion override them.

And finally, the developing embryo does not at first display the *structural* differences which characterize the male and female body in the adult. Moreover, *some* of these differences may be remarkably late in developing. This is sometimes termed *paedomorphism*, since it amounts to an embryonic stage which persists till later in life than is normal. In which case, it could have been that Adam was physiologically paedomorphic since he was presumably created as an adult, not as an embryo. I'm not speaking of his moral or intellectual development, only of his physiology before Eve was separated from him.

But let me repeat a previous observation. We *all* of us, individually, *begin* where Adam began. The only difference is that by the time of birth, a male child is manifestly male and a female child manifestly female. Yet both, even in the adult stage, retain certain features in their constitution which seem to be more characteristic of the opposite sex. In old age these features sometimes find expression: in women as facial hair, baldness, deepening of the voice, etc., while on more than one occasion men have successfully suckled a child when the mother died in childbirth. This is known technically as *gynecomastia*.

If we had seem Adam *after the divine surgery* we would not distinguish him from today's male figure, even on the dissecting table: but if we had seen him *as he first came from the hand of God* we might well have observed some very significant differences. Though, unlike us, he was created in an adult form, he may well have begun as we all begin – sexually neutral and capable of developing either way.

It is perhaps worthy of comment that the concept of a male/female (androgynous) nature in Adam's constitution as created was widely held by Jewish commentators in pre-Christian times and by some early Christian commentators under their probable influence. Pagan traditions show clear evidences of a similar view, though unlike the Jewish traditions they are filled with absurdities and are far less matter-of-fact.

The biblical account itself, of what took place during these crucial events, is a miracle of literary condensation. It does not tell us all that we might like to know, but nothing can be ignored in the record without destroying the meaningfulness of the whole. It will bear microscopic examination and yet the account is simplicity itself. This very simplicity enlightens the naive but confuses the worldly wise. The *words* are for children but the *thoughts* are for men, written as it were "for all sorts and conditions" that they may understand the truth at their own level. The record is agelessly up-to-date.

In order to explain the appearance of such a creature as the First Adam was, it is quite pointless, if not patently absurd, to appeal to any evolutionary process. Because what amounts to the division of the sexes and the *initiation* of sexual dimorphism is considered by the evolutionists to have already been a *fait accomplis* millions of years before the appearance of man who merely inherited it.

Thus it becomes apparent that we must recognize a strictly biological aspect in any theology of man and his redemption. There is a very complex physiological undergirding to the account of the formation of Eve out of Adam that is not merely intriguing but is essential to the whole working out of the Plan of Salvation.

There can be only *one* creative act, the creation of Adam. Every other human being, *including Eve*, must be a derivation from this one Federal Head of the human family. If Eve was not "in Adam", and therefore did not originate out of Adam, Adam was not her generic Head and the Lord Jesus Christ could not be her Saviour.

To refuse to recognize this fundamental fact is to undermine the very foundation of Christian theology in its strictly logical coherence.

Chapter 6

DEATH: A FACT OF LIFE?

Natural or un-natural?

Almost anyone "in the pew" if asked what kind of death Adam introduced by his disobedience, would say *spiritual* death. Theologians are pretty well agreed that because of Adam's disobedience, *physical* death *also* was introduced. But even they, all too often, play down the physical aspects of the Fall as though physical death is as natural to man as it seems to be for the rest of nature. What was threatened as a punishment, many would say, was not death per se but *premature* death, a death sooner than expected.

There is another group of Christian writers whose orientation tends to be more scientific than theological, who however take Genesis to mean that the cause of physical death as a penalty of man's sin applies not only to man but even to the animal and plant world.

And then, of course, there are the evolutionists in general who would not attach any factor of *penalty* to the introduction of death but simply say that death is common to all life and not related to the disobedience of man at all.

Who is right?

How universal *is* physical death? What actually *is* the

physiological cause? Does death in other organisms than man really occur for the *same* reason that it occurs in man, and is it the same phenomenon?

These are the questions to be addressed in this chapter. They will be dealt with very briefly, but hopefully looked at from all the basic angles which are or have been under scientific investigation in recent years. As a matter of policy, I am omitting all the usual paraphernalia of scholarship such as footnotes, documentation, graphs and charts: but every statement can be documented. For the reader who might be most likely to want such things, they will probably not be necessary since virtually every statement will be found widely acknowledged in the literature. While those "in the pew" won't be concerned about it, those "in the know" won't need it!

The Greeks had an ideal: to die young and postpone it as long as possible! Dying *senile* was what the Greeks wished above all to avoid. So do we. For most people approaching old age, death is often feared less than the senility which is apt to precede it.

Perhaps the vast majority of animals, by contrast with man, neither anticipate nor experience this long slow decline but retain remarkable vigour and apparent contentment until very near the end. A few domesticated animals may "die away" more slowly but chiefly because they have lived longer than they would have done in the wild so that the situation is un-natural.

The hazards of normal animal existence tend to be such that any significant decline in strength and vitality is quickly taken advantage of by other competitors and death is hastened for the individual in one way or another as a consequence. Man is exceptional in this slower pace of dying, and we shall have occasion to comment later on the significance of this fact. At the present moment my object is to look at the subject of physical death as a simple fact of life without regard to theological implications.

It should, throughout the discussion that follows, be kept in mind constantly that death is *not* the inevitable consequence of having lived. Some of the lower forms of life, and not a few plant forms (trees, for example), can continue their existence for thousands of years without any decline in vigour. There are some trees now known from Japan which may be 7000 years old! And many of the amoebae that share our world today shared the world of Adam. Moreover, there are some cells, cancer cells for example, that give all the appearances of possessing immortality. Death in itself is not a concomitant of life in itself.

It seems proper to start with a definition of death itself, and here we run into difficulty. In the first place, it is difficult to lay down a precise defintion of death without first of all having a precise definition of life – and this we do not yet have. One would think it would be a simple matter to define life but it isn't. There is a group of people who have committed themselves extensively to the view that there is no difference between what is living and what is non-living.

In all seriousness they go even so far as to say that if cells are alive then the components of the cells are alive: and if the components are alive, then the atoms which make up their substance are alive: and if the atoms are alive then even the particles – the protons and the electrons and all the rest of them – must be alive. The determination of the purely materialistic philosophers to avoid any hiatus in the transition from non-life to life drives them relentlessly to this seemingly logical conclusion.

Someone in exasperation suggested that you can tell whether something is alive or not, just by kicking it. If its response is predictable, given sufficient background information, then you can assume it is dead: if it is quite unpredictable, then you can presume it is alive!

Now, at the present stage of medical wizardry it has

suddenly become very important to be able to determine when a human being is dead or merely in deep coma. Individuals declared to be clinically "dead" have more than once in recent years shocked surgeons about to remove their kidneys for transplant purposes by suddenly sitting up on the operating table and asking them what in the world they were up to. Furthermore, certification of death has become more and more difficult in view of the currently available means of sustaining life artificially. The case of Karen Quinlan is a distressing illustration of this fact.

In 1975 Karen, then a teenager, swallowed a murderous cocktail of alcohol and tranquilizers. She has been unconscious ever since. She is now an insentient, motionless, curled-up, skin-wrapped skeleton, having "recovered" from the cocktail to the extent that she now survives without a respirator.

No one even pretends to believe that she will ever recover consciousness, while in the meantime the bill for this cruel exercise has reputedly passed the three million dollar mark. "She" (if this preparation can be personal still) has effectively been condemned to *life*, not to death . . . . a new thing in human history. And no one has the right (it seems) or the courage (perhaps) "to blow out the candle" (if the candle is alight) in order to allow this tragic example of man's over-sophistication to be terminated. In Karen Quinlan's case, how do you define life? And how do you define death?

But some very serious attempts have been made in recent years to define what death is. In 1968 the Harvard Medical School tried to give an answer in a statement issued under the heading, "A Definition of Irreversible Coma" in which were listed four basic criteria:

> 1. *Total unresponsitivity*, i.e., total unawareness of externally applied stimuli, even when painful, judged by vocal or other forms of response such as groaning, limb withdrawal, or quickened respiration.
>
> 2. *Total absence of movement* over a period of at least one

hour. This would include detectable pulse or respiration. It recommended artificial respiration be cut off to see whether any attempt at breathing would be made within three minutes.

3. *No reflexes*: the pupil fixed, dilated and unresponsive to movement or variation in intensity of a light source. Since the establishment of a fixed dilated pupil is clear-cut in clinical practice, it was felt that there should be no uncertainty in such a case.

4. *Flat encephalogram* (= 'cerebral silence') for twenty-four hours with no measurable change.

In each case the assumption is made that there is no evidence of hypothermia (a body temperature below 90° F. or 32.2° C.) or the use of central nervous system depressants such as barbiturates.

But how this definition of death fits the case of Karen Quinlan is difficult to see since breathing continues to be observed in the absence of artificial assistance. Karen would appear to be alive in so far as purely physiological considerations are concerned. But apart from these borderline cases which keep cropping up, there is no doubt that unequivocal death comes to millions of organisms, even if we have difficulty defining it.

This is not the time or place to go into the distinctions that now have to be made between *necrosis* (the death of cells locally but not the death of the whole organism); *clinical* death when the doctor says the patient is dead; *legal* death when the coroner clears things for the undertaker to proceed and the will to be probated; and *biological* death which signifies that heroic measures to resuscitate are futile and the body is already disintegrating.

But to complicate the matter even further, in both man and animals death can be viewed as either a process or an event. It is not certain whether (apart from accident) it is ever truly an *event*. When God told Adam that "dying he would die", we have a hint of the fact that dying is a *process* and many now believe that we are dying from the moment of birth. The same may be true of animals, though not on account of sin but for purely physiological reasons which may perhaps be 'natural' for the animal because it suggests a mechanism which is set to program the animal's life span in order to prevent over-population.

But there is almost always a terminal period in which what may have proceeded for a long time at a slow rate suddenly gallops away, and life is brought abruptly to an end.

The evolutionists, while admitting that death is not a concomitant of life, are driven to argue that it must have been invented and preserved by nature because it is both necessary and beneficial. Nature is assumed never to invent except on a utilitarian basis. It is held to be necessary for the following basic reasons:

1. To prevent overpopulation by any one species.
2. To leave the way open for further evolutionary progress by ensuring the removal of less successful species which would otherwise clog the system.
3. To remove diseased, malformed, or less vigourous members of a population so that they do not perpetuate themselves. The vigour of the stock is thus preserved.
4. To provide food for predators of a higher order which by a carnivorous diet have a higher energy level, such a higher level of energy being considered a superior form of life.

Let us consider these in slightly greater detail.

1. It is obvious that some species, rabbits for example, breed much more rapidly than, say, stoats. In the economy of nature it is logical that a world over-crowded by a single species could, under certain conditions, become fatally infected with a disease peculiar to that species which could wipe them out entirely, and the web of life accordingly might be grossly disrupted. Or such a prolific species could consume all vegetable foods necessary for other species to live on, and so bring about their extinction. Planting only a single species of tree in reforestation may be unwise be-

cause of the danger of some species-specific disease wiping out the whole area and robbing the soil of ground cover.

By the same token, it would not do to have stoats multiply explosively without constraint, for an opposite reason. They could endanger every other species because of their efficiency as predators.

Professor Edward O. Dodson gives a striking illustration of what could happen but for death in the case of starfishes. In one species the female spawns some 1,000,000 eggs in a year. In a small area of only a few square yards there may be 50 such females. Each year the eggs laid would therefore be 50,000,000. Assuming 25,000,000 of these are females, at an ordinary rate of reproduction these 25,000,000 females would produce 25,000,000,000,000 eggs. It would take only 17 years for the number of starfishes on the earth to exceed the total number of electrons in the whole visible universe!

2. In the long view, evolutionists hold that in any species there will be individuals born with some slightly modified structure or habit or instinct which has the potential for real advantage to the species. But at first the numbers of such individuals will be too small for the advantage to be shared widely. It could also happen that the *potential* advantage is an *actual* disadvantage until the number of individuals who possess it mark out an niche for themselves and multiply sufficiently.

The competition for food and mates is assumed to be keen. Thus if the more dominant mates of the old stock are not somehow removed, the new stock will not be established in time to preserve its unique and advantageous characteristics. But if the old stock have limitations placed upon their life expectancy, the newcomers' chance of survival and multiplication will be enhanced.

If we transfer this scenario into the business world, we see something of the basis for fixing mandatory retirement age at a lower rather than a higher level. Extending the term of office of senior staff can work to the disadvantage of a company by stifling the initiative of the younger members who see little hope of advancement.

3. This situation is illustrated very nicely by the finding in Northern Canada some years ago, a finding since repeated elsewhere, that when the wolf population is reduced by man to below a certain level with the intent of increasing the deer population, the opposite effect may occur.

The reason for this is now believed to be because when the predatory wolves are reduced in number, the deer population does indeed increase but the increase includes sick or less fit deer that would otherwise have been killed off by the wolves. The well-being and vitality of the deer herd declines as a result and they begin to die off at a higher rate than expected. Matings are less successful and the offspring less healthy.

By leaving the wolf population to set its own level, the deer herds are healthier and prove better able to survive the vicissitudes of northern weather.

4. In the animal world, survival depends to some extent upon individual energy levels. Animals which eat meat can maintain higher levels of energy with less time spent foraging for food than comparable herbivorous species.

It is a known fact that a man in good condition can run down a horse. It is obvious that the horse can run faster than the man, but the horse, being herbivorous, tires more quickly and will have to keep stopping to browse. Each time the man begins to catch up, the horse will be forced to run again. In the end the horse exhausts its energies and allows itself to be caught. The Indians of the Plains could apparently catch horses this way, and such horses once caught were less likely to run away even when free to do so.

There is a case of a man who is said to have volunteered to obtain medicine in an emergency for Queen Elizabeth I. The distance he had to go was some thirty to forty miles, and a man was sent out on horseback at the same time in case of accident. The runner made the round trip of over sixty miles in less time than the horseman, and he was interestingly awarded by the Queen. He received a new suit of clothes to compensate for the spoiling of his old suit on the trip, the award to be repeated annually: and the head of his household in each succeeding generation was promised the same reward each year in perpetuity. Only Cromwell's Republic put an end to the arrangement, so it is said, or perhaps his descendants would still be beneficiaries.

Whether the story is true or not, it is not difficult to see that energy is more quickly recovered on a pre-digested diet of meat than on vegetables, since one step in the conversion to energy has already been taken. Death by predation thus advances the course of evolution, according to this thesis.

As to the *mechanism* of death, there have been scores of explanations. A few have won wide acceptance for a while, but fewer still have stood the test of time to the present. The following are six more or less current explanations of ways in which death may occur 'naturally' – apart from accident or predation.

1. Animals often display an internal mechanism which "self-destructs" the organism once it has become infertile.
2. A built-in limitation to cell doubling and therefore to new growth and tissue repair, has been observed. The life span of the whole organism is therefore believed to be "programmed" accordingly.
3. "Wear and tear" brings about reduced viability so that the animal succumbs to stress.
4. DNA errors of transcription accumulate until a point of "error catastrophe" is reached.
5. Since millions of sperm are released by the male and only a few can continue their existence by fusion with the ova, millions are destined to die.

6. Unicellular animals that multiply by simple division appear to escape natural death. It is concluded that procreated animals are mortalized by some factor relating to or associated with, the process of gestation and birth.

Let us elaborate these alternatives briefly.

1. According to Jerome Wadinsky of Brandeis University, this "self-destruct" mechanism is found in a number of species of which an excellent example is the octopus. When fertility is at an end, the individual animal under some hormonal influences ceases to eat and dies of starvation. By experimentally preventing the hormonal effect, it is found that the animal will continue to eat normally though remaining infertile, and its life is considerably prolonged.

The evolutionary explanation is that from the point of view of the species (as opposed to the individual), an animal no longer fertile will make no further contribution to the improvement of the species but will still continue to compete for its food resources. A mechanism has thus been invented by nature to save this waste of food by putting an end to the individual. Man is one of the few creatures that continues to live for a comparatively long time after ceasing to be fertile.

2. Leonard Hayflick, over a period of years and by some very elegant experimental techniques has proved to the satisfaction of many biologists (though by no means all) that cells have a limited capacity to divide and multiply. The organism thus loses the ability to continue growth and to heal wounds.

Since growth and continued life appear to be almost synonymous, a cell that has the capability of only a certain number of doublings (one cell becoming two, two cells becoming four, and so on) the organism has what is termed a "spanned" life which it rarely exceeds significantly. Different species have either a different rate of cell division or a different allotment of doublings.

The finding not only limits the life span of the animal but probably also its size. Since there are real limitations to the size of any free-standing animal, this seems to reflect a certain wisdom in nature. Some years ago, J. B. S. Haldane wrote a fascinating paper on the subject which he titled "On Being the Right Size"! A neat balance is thus struck between the ability to heal, which requires a high level of cell multiplication, and growth in size which must be controlled. It also effectively spans the life of the individual organism.

3. The wear and tear theory of the cause of death is a common-sense view since it seems obvious, but it is no longer taken seriously because experimental evidence has shown that stressed animals (whether physiologically or psychologically) live just as long, and sometimes longer, than unstressed animals. Records of hard-worked animals in captivity, like elephants and horses, by comparison with free animals in the wild show, according to Professor Raymond Pearl, that the worked animals out-live the wild animals, contrary to expectation.

4. Mutations, or DNA transcription errors, are constantly occurring in animal cells for a number of reasons not yet fully understood. If a sufficient number occur in a given cell, it will cease to function. If a sufficient number of non-functioning cells occur in a particular organ, it will fail as an organ. At some point these errors overwhelm the whole animal and what Dr. Leslie Orgel of the Salk Institute has aptly termed an 'error catastrophe' occurs which results in the death of the whole organism.

At the present time this appears to be the most acceptable view of the *mechanism* of animal death.

5. According to Professor B. Bacetti of the University of Siena, there is evidence that in a number of species, the extraordinary proliferation of spermatozoa of which the

vast proportion die without contributing to the fertilization of the ovum, serve the purpose of providing by their decay a protein-rich environment for those which succeed. This is true of certain species of gastropods.

A rather similar situation exists in the case of codfish. The female spawns some 6,000,000 eggs of which only 4 or 5 survive. It is believed that the non-surviving ova, in a like manner, provide a protein-rich environment for the eggs which do survive to grow more vigourously.

6. August Weismann in the late nineteenth century was the first to underscore the fact that asexual reproduction leads to potentially immortal animals whereas all sexually propagated offspring appear without exception to be mortal.

It is not clear why this is so in nature, but it does suggest that if an animal wants to live for ever it should avoid being born!

There are a number of people who feel that the death of animals is a direct result of the Fall, and they occasionally point to Romans 8:22 as implying this. Here we are told that "the whole creation groans" as man awaits his redemption. But the fact is that the phrase "the whole creation" is recurrent in the New Testament and clearly applies to human kind, not to animals.

In Mark 16:15 the same phrase in the original Greek surely cannot mean that the Gospel is to be preached to *animals*, as Francis of Assisi preached to birds. Colossians 1:15 (again the same phrase in Greek) cannot mean that the Lord Jesus Christ was the firstborn of all the *animals*. In Colossians 1:23 (once more the same phrase) Paul did not mean that the Gospel had already been preached to *animals* everywhere!

And finally, Romans 5:12 which speaks of the entrance of death – or as Martyn Lloyd-Jones acutely observes, the "invasion" of death – cannot be extended to animals as some would have it to be, because the proof of this invasion according to the verse itself is that all have sinned. The universality of death for man is proven by the universality of man's sin. The whole point of the passage is that *all men* die because *all men* are sinners. Animal and plant life is clearly not in view. Moreover, amoebae and paramecia are animals; yet they are not subject to death except by accident. That such creatures are tiny has no bearing: they are alive. Thus the concept of death as universal and due entirely to man's sin is neither supported by the facts nor required by Scripture.

It is clear that death exists in nature not as a punishment but ultimately for its maintenance and well-being. Therefore it is most important to realize that none of what has been said about the necessity of death in the animal and plant world applies to man. *This cannot be emphasized too strongly*. For them death is natural: for man it is not.

But it is also evident from what we know about amoebae and paramecia (and unfortunately, cancer cells) that cell life is not necessarily subject to death at all. Once created, it would appear to have potentially the capability of endless existence, barring accidents. The phenomenon of life per se thus illustrates Augustine's concept of immortality as something which it is not impossible to kill but which otherwise has no necessity of dying.

The death of man and the death of animals are thus two very different *kinds* of death. In the case of animals, death is a necessity, something programmed for them for the benefit of the web of nature as a whole. No single species of animal is allowed to multiply and "fill the earth", though taking all the species together they have in sum successfully fulfilled the command of Genesis 1:22 to fully occupy it.

With man, the case is different because, though he is a single species it was intended that he should fill the earth and govern it, but there is good reason to suppose that each individual, as soon as he or she had matured, would have been removed from the earth by translation without tasting death at all, thus preventing over-population. No such removal by translation was planned for any of the other species. Only man was designed for this happy prospect.

Thus the *why* of death for animals has no bearing on the *why* of death for man, even though the mechanism may not be altogether dissimilar since both live rooted in the same natural order. But for animals, death being appointed like all else in nature, makes it natural: whereas for man, though death is also now appointed (Heb.9:27), it was appointed only as a penalty and as such was *un-natural.*

Therefore it seems clear that Adam's un-natural death is no more to be accounted for by some supposed animal ancestry, than the natural death of animals is to be attributed to the Fall of Adam.

Chapter 7

THE CONSTITUTION OF MAN: WHERE ONE + ONE = ONE

A Human Spirit + A Human Body = A Human Being

It has often been observed that the Hebrew people were intensely religious by inclination but never felt any need to structure their faith or systematize it as a theology. Of commentaries they wrote many, and the Talmud grew apace year by year. But despite its great volume of traditional law, it contained little that could qualify as theology in the Gentile sense. Strict adherence to logical systematization of their beliefs did not seem to interest them, though they did systematize their *practices*.

This still seems to be essentially true. A recent edition of *The Standard Jewish Encyclopedia* which runs to very nearly 2000 pages, under the heading **Theology** simply says: See *God, Judaism, etc.* The entries under *God* and *Judaism* bear little resemblance to our theologies, being more history than anything. And what the *et cetera* means is hard to say, since obviously one cannot find it anywhere in the Encyclopedia!

Although in the Gentile world we have, since the earliest days of Christianity, continually produced and refined Creeds, Definitions, and Statements of Faith, we have not founded the substance of these formulations on the Old Testament but on the New, especially upon Paul's Epistles. The fact is that Hebrew disinterest in such formulations is reflected in their Scriptures.

And it should be borne in mind that these Old Testament Scriptures strictly include part of the New Testament, namely, the Gospels. Because, *until Calvary*, the statutes and ordinances and rituals, and the Abrahamic promises to Israel which relate to earthly matters, were all still in force.

The consequence of this is that when we try to construct a biblical theology (or psychology?) of the constitution of man, we find little precision in the Old Testament or in the Gospel records. Precision belongs chiefly to Paul.

One can appeal to passages in the Old Testament that often strongly support a New Testament theology, but they are also sometimes contradicted by other passages in the Old Testament and are thus of only slender evidential value. This brief chapter is therefore essentially a New Testament construct, but it is both revealing and satisfying in its simplicity. And as will be noted later by reference to Barton Payne's writings, there is a comfortable agreement between the Old and the New Testaments in this biblical anthropology even though it could not have been constructed on an Old Testament basis alone.

There has always been a debate as to whether man is composed of *two* distinct components – a body which is physical and a spirit or soul which is not – or whether man is composed of *three* distinct components as seems clearly to be implied in 1 Thessalonians 5:23. The first formulation is referred to as a *dichotomy* and the second as a *trichotomy*.

Sometimes the *three* component advocates (trichotomists) feel they have support also from Hebrews 4:12. However, if one wants to insist on a strict literalism in this passage, one could argue for *four* components, made up of soul, spirit, joints, and marrow: which would, I suppose, be a *quadrichotomy*. Interestingly, some versions reduce this quadrichotomy not merely to a trichotomy by counting joints and marrow as one element, but to a dichotomy by taking soul/spirit as one component or simply what is non-physical, and joints/marrow as the equivalent of flesh and bone or simply what is physical. They thus argue that the writer is saying that the Holy Spirit is able to set the spirit against the body.

The situation is complicated by the fact that in addressing a Jewish audience (as the Epistle to the Hebrews does), the Lord Himself used a number of terms, each of which might be taken as a separate component of man's constitution: *strength, spirit, soul, heart, and mind* (cf. Matt.22:37 and Mark 12:30). This could, I suppose, be called a *quinquichotomy*!

Really, we are left with only one passage clearly contending for a trichotomy (1 Thess.5:23), the rest of the New Testament strongly suggesting that man is simply a dichotomy of body and spirit. And for purposes of ordinary discussion, few will challenge the convenience of being able to view man as an embodied spirit, a created spirit in a procreated body. When the God-given spirit is infused into the body, the soul emerges. As the simplest of all equations, spirit + body = soul, for the soul is the person, *the whole man*, the self. For you it is "your self", for me it is "my self", body and spirit.

Let us look, then, at some of the evidence that the New Testament (especially the epistles of Paul) almost always speaks of the non-physical component of man as *spirit* not *soul*, inspite of popular opinion to the contrary. Whenever we find a clear reference to man as a dichotomy, it is always as a dichotomy of body and spirit, not a dichotomy of body and soul.

James 2:26 tells us that the body is dead without the *spirit*. No hint of departure of the *soul* is present in this simple observation. Jesus in his 'departing' commended not his soul but his *spirit* into the Father's care (Lu.23:46), as did also Stephen (Acts 7:59).

Accordingly when resuscitation takes place, it is not the soul but the *spirit* that rejoins the body (Luke 8:55 and Rev.11:11). And this is also true in Ezekiel 37:5, 8-10 where the word for *spirit* in Hebrew is here, however, rendered 'breath' in the King James Version, although there is a perfectly good word in Hebrew for what we mean by *breath* namely, *neshamah* – a word which was not used in the original though it could have been had this been the intention of the Author.

It is significant that John 3:6 speaks of the re-birth of the *spirit,* not the soul. And in 1 Corinthians 5:5 we are told of the saving of the *spirit* rather than of the saving of the soul. In Ecclesiastes 12:7 it is the *spirit* not the soul that is given to the new-born, after being presumably preformed by God (Zech.12:1). And it is the *spirit* not the soul that is surrendered by Ananias and Sapphira (Acts 5:5,10); it is the *spirit* not the soul that is willing though the body is weak (Matt.26:41); that is finally to be made perfect (Heb.12:23); and that cannot be retained by man when the time comes to surrender it back to God (Eccl.8:8).

In all these important passages, and often contrary to the way they are quoted, it is the spirit and not the soul that is spoken of. We speak easily of the saving of the soul. And while this is perfectly justified, as will be seen later, it is not strictly biblical. Passages where spirit is used instead of soul can be multiplied greatly by careful attention to the wording of Scripture. Thus it is both body and *spirit* that need cleansing (2 Cor.7:1). Mystically, the Church is one body and one *spirit* (Eph.4:4). And we are called upon to glorify God in our *spirit* and our body (1 Cor.6:20).

What, then, of the *soul*? Where does it enter the picture? Surely, the soul is the end result of the fusion of body and spirit, an entity, a reality, generated by the fusion of two elements – just as salt is generated from sodium and chlorine gas, or as the colour green is generated by the fusion of yellow and blue. When the two components are separated by death, one component returns to the earth and the other to heaven into God's keeping, until reunion of the two at the raising of the body brings about the reconstitution of the person. The soul is the result of a union, an entity which comprehends the whole man.

Such a view can be supported both from the Old and the New Testaments. J. Barton Payne has written a most useful volume entitled *The Theology of the Older Testament*. In his chapter on "The Nature of Man", he proposes that in the Old Testament we have the following progressive equation:

DUST + BREATH = FLESH (as the living organism)
and FLESH + SPIRIT = SOUL (Heb. *nephesh* i.e., the person)

The simplicity of this arrangement commends itself highly since it seems to meet many of the apparently conflicting relevant passages in the Old Testament, and virtually *all* of the New. Passages not satisfied are usually poetry or analogy or an accomodation to common parlance. Basically it presents us with a physical body and a non-physical spirit which together constitute the soul, the person.

We can compare with this a recent equally satisfying study of the biblical view of human constitution by Robert H. Gundry titled *Sōma in Biblical Theology*. This makes an excellent companion volume on the New Testament evidence to that by Barton Payne on the Old.

Gundry concludes, on the basis of the New Testament, that man is a body plus spirit entity which, when fused, becomes a soul. Notice that we are talking about fusion, not mere addition. Soul is something which neither body nor spirit alone can ever be. Man thus can be said to *have* a body and to *have* a spirit: but he does not *have* a soul. Man

IS a soul.

Without the spirit the body is like a car without a driver: without the body the spirit is like a driver without a car. A brief quotation from Gundry nicely indicates his overall position:

> The biblical touchstone for truly human life is not consciousness of the spirit, let alone the material being of a physical object such as the body. Rather, man is fully himself only in the unity of his body and spirit in order that the body may be animated and the spirit may express itself in obedience to God.
>
> Both parts of the human constitution share in the dignity of the divine image. That dignity lies in man's service to God as a representative caretaker over the material creation. For such a task man needs a physical medium of action as much as an incorporeal source for the conscious willing of action.
>
> Neither spirit nor body gains precedence over the other. Each gains in union with the other: each loses in separation from the other.

We seem therefore to be nearest to the truth when we formulate the simplest equation possible:

BODY + SPIRIT = SOUL.

This does not really make man a duality, except for purposes of analysis and discussion. Man does not exist as a person when body and spirit are separated, and therefore it is only in a manner of speaking that we can talk about the body as half the man and the spirit as the other half, since there is no such thing as half a man. When separated, the body at once ceases to *be* a body and becomes merely a purposeless conglomerate of chemicals: and the spirit appears to lose all contact with physical reality and all means of expression. Consciousness is almost certainly lost.

Only the resurrection of the body and its re-fusion with the spirit can reconstitute the whole man, the person, the soul. A SOUL is therefore a *monad*, that is, an absolute singular reality, by nature indivisible. It comes into being as an entirely "new thing" as a consequence of the fusion of two elements. It is thus apparent that we can restate the equation above in a new form:

ONE + ONE = ONE.

Spelled out, such an equation means that "one and one makes one", not TWO: and the secret of the resolution of such an odd equation then lies in the meaning of the little word "*and*". What do we mean by *and*? In this case, not merely *plus* but "fused with", "made one with" in the most literal sense.

We are by no means without scriptural analogies for this form of equation. The most obvious one is to be found in Genesis 2:24. When God brought Eve to Adam (the Father bringing the bride to the groom!), He said, "They shall be one flesh". One plus one equals one.

We find the analogy again in the unification of the Body and its members (i.e., the Church) and the Head (which is Christ) becoming a single functioning organic unity. "For as the body is one and hath many members, and all the members of that one body, being many, are one body; *so also is Christ* . . . . ye are the body of Christ . . . . and He is the head of the body" (1 Cor.12:12 and 27, and Col.1:18). Likewise, Jew and Gentile are so joined as to "make of the two *one* new man" (Eph.2:15).

Thus though it takes two partners to MAKE the "marriage", it is not the partners who ARE the marriage. The marriage which is thus generated by the partners becomes a reality all of its own. As yellow and blue MAKE green, neither the yellow or the blue by themselves ARE green.

Side by side they remain yellow and blue: fused they become green.

In an analogous way, the body and the spirit do indeed MAKE the soul, but it is the soul that is *made*, which in this analogy is the MARRIAGE, not one of the partners. It exists in its own right, just as the resulting soul exists in its own right.

An evil spirit may take over a human body, or even a dead one – as Satan tried to do with the body of Moses (Jude 9). But this does not generate another human *soul*, nor would it have done so if Satan had succeeded. In the latter instance, it would have generated only a monster. What constitutes a human soul is the unification of a human body with a human spirit.

In terms of his constitution, man is a body and a spirit: in terms of his soul, man is "simple and indivisible" as the theologians have it. The soul of man cannot be divided and survive as a soul. If the two components are separated, which is the only division that can be made, the soul no longer exists. What God planned was not just the components of man by which He created him in two stages, first the body and then the spirit. When the spirit was infused into the body which had been prepared to receive it, then man "became a living soul" (Gen.2:7). *The living soul* is what God created "in his own image".

The union of the spirit and the body is a "marriage by appointment" since each spirit is created specifically for each body, and it was never God's intention that this union should be dissolved in death. Although through sin it will be dissolved, reunion will occur when the resurrected body is joined by the spirit. It is not our "hope of glory" to be merely a redeemed ghost but to be a redeemed soul.

Even this is not a sufficient statement, because it suggests a spirit consciously in search of its body. I believe there is no such thing as a *conscious spirit* without a body. It needs the body's brain to have consciousness of the real world and even of itself. Moreover, I am confident that, for the Christian, there is no lapse into unconsciousness when we pass into the presence of the Lord, for our spirit is instantly rejoined to its resurrected body. In short, death becomes our resurrection.

The *soul*, therefore, is in the strictest sense indivisible, for the only division that can be made results in the dissolution of the soul. Meanwhile the spirit that has departed from the body passes directly into God's keeping until the body is resurrected to form its proper home.

Between the departure of the spirit and the resurrection of the new body, however, there is no experienced lapse of time because there is no "time" to lapse. Thus we are *never*, at any stage in the life of the world to come, reduced to the status of a mere ghost. We pass into the presence of the Lord clothed, not naked (2 Cor.5:4), and to be rid of this body is to be clothed in a new body suited to a royal reception.

I have spelled this out very carefully and fully in my book *Journey Out of Time*, a book which sheds a new light on a number of passages of Scripture that bear directly upon the circumstances in which this journey into eternity will be made. The Lord's promise that the believer will not "taste" of death (John 8:52) will be literally and wonderfully fulfilled. There will be no experienced loss of consciousness when this journey is made.

So what we really have is this. A human being is by definition a human spirit fused with a human body, not a mere combination of spirit and body but a fusion in the most absolute sense.

In view of this equation, it is therefore quite proper to speak of the saving of the soul – i.e., the whole person – because with every assurance of the rebirth of the spirit and every assurance of the redemption of the body, we do indeed have, effectively, every assurance of the saving of the soul. This is not the saving of half the man, but the saving of the *whole*.

The evolutionary view of the origin of the soul has to be, in the final analysis, a kind of spin-off or epiphenomenon that emerges directly from the material of the body and will last only as long as the body lasts. It is a philosophy which holds out no transcendental value for the human spirit so long as its own interpretive canons are faithfully applied to the evidence it admits. The Christian evolutionist appears to me either inconsistent in his theology as a Christian, or inconsistent in his philosophy as a Scientist.

A Footnote to this Chapter.

The reader who prefers to view man's spiritual nature as composed of more than one element (heart and mind, or soul and spirit, or mind and will) need not feel offended. The only point that I feel is quite crucial to my thesis is that **man is not man at all without his body.** *His spirit is, in fact, not more important to his future life hereafter than his body is.*

Man is a dichotomy in the sense that he is composed of a non-physical part of his being which is his spirit and a physical part of his being which is his body, and he cannot be whole without the total fusion of the two. The resurrection of his body is every whit as important to his future identity as the preservation of his spirit. Together they constitute the survival of his person.

Chapter 8

HOW DID I COME TO BE ME?

The Origin of the Human Spirit

We know that we receive our body from our parents by an act of procreation: but whence comes the other half, the spirit.

There are really only four views: (1) by Reincarnation, (2) by direct Creation, (3) by inheritance from our parents much as we receive our bodies, termed Traducianism, and (4) by Evolution. We will examine each of these very briefly noting something of what is the meaning and evidence for each, and some of the problems each view creates.

(1) By REINCARNATION. Broadly speaking, reincarnation means that a soul passes through a succession of bodies, each of which becomes a temporary means for the expression of its condition until perfection is reached by experience. At this point the soul either passes into the total rest of immersion in the sum of "cosmic consciousness" and is finally freed from the burden of personal identity, or becomes one with God – perhaps without loss of that personal identity. Many people find the cost of individualism is too high and prefer to be "lost in the

crowd". Final absorption in something much bigger than self seems a very happy solution to such people.

The Jewish people gave considerable thought to the idea of reincarnation and crystallized their ideas about it during the Middle Ages in a work referred to as *The Cabala* or *Kabbala*. Their thinking was strongly influenced by gnosticism which viewed embodiment as degrading to the spirit and reincarnation undesirable accordingly.

In this view, all souls were created at the beginning and perfectly content without bodies. Such souls were androgynous by nature. When subsequently, as a punishment, souls were embodied, each was divided into male or female so that they now seek reunification with themselves by marriage. Since embodiment was a penalty, a strong leaning towards asceticism naturally developed.

Nicodemus probably had in the back of his mind some kind of reincarnation when he asked the Lord whether a man could enter into his mother's womb a second time and be re-born (John 3:4).

The possibility that John the Baptist was a reincarnation of Elijah or Jeremiah or one of the prophets (Matt.16:14) is another illustration that the idea was quite familiar to them, though not yet logically formulated.

Later, Origen (185 - 254) was to develop the concept much more fully and subsequently to be roundly condemned as a heretic for his pains by the Western Church. However, he still continued to be recognized as the first of the systematic theologians and as an intensely creative writer. His works became a veritable watershed of original ideas.

As to the evidence, it was then what it still is, anecdotal in nature. It is a common experience for many people at one time or another to come upon a scene or to meet a stranger, and experience a strong but odd feeling of having been there before (*déjà vu*) or having met previously. There are a few cases on record of individuals who described a place in great detail as though it were quite familiar which they had not however actually visited: and their description has proved to be remarkably accurate.

Those within the framework of Judaeo-Christian philosophy, have limited reincarnation to the human *spirit*. Other religious communities outside of this tradition have been quite willing to concede that man might in the next reincarnation be an animal or an insect or even a plant, and correspondingly any animal might be reincarnated as a human being. Origen felt that the problem of man's sinful nature was best explained by some prior existence in a less than perfect state.

The concept of reincarnation really provides no solution to the origin of the soul or spirit. It merely proposes a history of what happens to the soul *once created*. We are still left with the problem of where the soul originated and in what form.

Is there some kind of reservoir which, like water, can be fragmented into droplets and allowed to fall into bodies as they are appropriated? Or is there a store of individually created souls predestined to fit each body when the time is ripe? And in that case, in view of the steady increase in the world's population and the number of bodies needing ensoulment, are there multitudes of souls on a kind of waiting list? Or are they created as needed?

Granted that with each death there is a soul free to serve somewhere else, the fact remains that births exceed deaths so that there are always more new vessels to be filled than old ones just vacated.

It seems that by and large the biblical data do not support the concept of reincarnation, especially since, for the redeemed at least, death does not leave a soul homeless but frees it for immediate union with its newly resurrected body.

(2) by Direct CREATION. This is the belief that the soul is, in each individual case, a direct work of God.

Many passages of Scripture can hardly be understood

in any other way. We have already referred to some of them in another context. It will not be amiss to repeat a few here to complete the record in this new context.

The spirit is *given by God*, not derived from man	Eccl.12:7
The spirit is *formed* within the individual *by God*	*Zech.12:1*
The spirit (or soul) is *made by God* (Heb.ʿasah)	Isa.57:16
God is the *Father* of all spirits	Heb.12:9

If it is God who has designed and created our spirit, suiting it to our body (whether male or female, introvert or extravert, practical or philosophical, artistic or unimaginative, creative or merely appreciative, etc.), it is quite natural that it should return to Him when it leaves this body. Here in God's hands it is preserved *in its identity*, to be infused into the new resurrected body which we receive at the instant of our relinquishing the present one. This resurrected body will match the character of its perfected spirit, thus entirely reconstituting a soul.

The spirit is only one half of our personal identity, and because it is of divine origin, the dissolution of the body which houses it in no way signifies that the spirit ceases to exist. It is a creation of God designed to last for ever, not an accidental and temporary by-product of a blind evolutionary process. God is able to reconstitute the whole man by the resurrection of his body in some identifiable form and by the re-infusion of the spirit into it. Such a destiny for man is entirely outside the purview or competence of evolutionary philosophy.

Logically, the evolutionists must of necessity reject *any* concept of direct creation of any essential component of man's constitution, such as his spirit: and especially any concept which demands continuous creation upon billions of occasions to meet the requirements of an ever growing world population.

(3) By TRADUCIANISM. This is the belief that the soul or spirit is procreated by the parents, along with the

body. The word *Traducianism* is derived from a Latin verb *traducere* which means "to transmit". The term was originated to convey the idea that in Adam and Eve a soul was directly created to form an inexhaustible reservoir of soul-stuff for their children. Derivation of all future souls did not diminish theirs, any more than the lighting of a second candle or a hundred candles from a first one diminishes the prime source. All "soul substance" was invested by one act of creation in Adam and Eve. Our souls are derived from that investment.

The great advantage of this concept is that it so nicely accounts for our inheritance of Adam's acquired sinful nature and guilt.

There are some problems which this otherwise attractive thesis raises.

First, there is no unequivocal biblical evidence for such an origin of the human soul or spirit. The only passage which might seem to qualify as such is John 3:6, "That which is born of flesh is flesh and that which is born of Spirit (or spirit?) is spirit".

If we use a lower case letter for *Spirit* to make it read as "that which is born of spirit is spirit", we might seem to have a strong basis for Traducianism. But the context clearly indicates that the word 'Spirit' has reference to the Holy Spirit and that the birth spoken of is a *re*-birth. "Marvel not that I said, Ye must be born *again*".

This has been recognized tacitly in the great majority of modern versions, including the RV, RSV, NASB, TEV, NIV, NKJV, Berkeley, Young, Williams, Jerusalem, Wuest, New American (RC), Fenton, Smith and Goodspeed, Rieu, and the Vulgate. Very few have used the lower case 's' (i.e., spirit). These include NEB, Rotherham, Barclay, and Phillips. Admittedly, majority opinion does not settle such matters, but it is noteworthy that when Adam first set eyes on Eve he did not exclaim, "This is now soul of my soul and flesh of my flesh" but "This is now bone of my bone and flesh of my flesh" (Gen.2:23). There is no suggestion in this that Eve derived her soul or spirit from Adam as well as her body.

Again, while God Himself is spoken of as "the Father of spirits" (Heb.12:9), Adam is never so described. Yet certainly such a title would have been particularly appropriate in his case if Traducianism had been true.

There is one further reason for rejecting the Traducianist position, namely, that it had found its way into the Early Church as a result of the influence of early Greek philosophers, chiefly Zeno (c. 450 BC) and Cleanthes (301-252 BC), both of the School of Stoics. Such a source for the doctrine would not commend itself to most students of Scripture since Greek philosophy is hardly a promising basis for a biblical theology.

Nevertheless, there were a number among the Early Church Fathers who did adopt it simply because it explained so nicely how one man's sinful nature could become universal by propagation, and not by example as Pelagius had argued. Tertullian was strongly in favour of it. To quote his words: "Our first parent contained within himself the undeveloped germ of all mankind, and his soul was the fountainhead of all souls; all varieties of individual human nature are but different modifications of that one spiritual essence. Therefore the whole of nature became corrupt in the original father of the race and so sinfulness is *propagated* together with *souls*". It is an attractive alternative, but not a biblical one.

However, if Traducianism is true and we assume our soul as having been received jointly from both parents, the Lord's soul was not a creation but was received from *both parents*, the two parents being the Holy Spirit and Mary. Thus, half of the Lord's soul is traceable to Mary, and though the other half is traceable to the Holy Spirit, the Marian half is corrupted. This half would corrupt the whole, and thus the absolute sinlessness of the Lord's spirit must be called in question *unless* we accept the Roman Catholic dogma of the Immaculate Conception which arose to deal with this very problem.

It has been suggested that Luther adopted Traducianism in order to place his position at one further remove from Roman Catholic doctrine. But the evidence seems to show that, like Augustine (who greatly influenced his thinking), even Luther himself was never completely settled in his own mind on the matter, wavering back and forth between Traducianism and Creationism. Those who followed him were clear enough and came out on the side of Traducianism, but Luther himself remained equivocal.

It is not unnatural that Christians who favour the evolution of man's body and therefore see no discontinuities in the processes of nature, tend to find Traducianism more acceptable since it, too, avoids even the discontinuity that direct creation of the *spirit* introduces.

4. By EVOLUTION. Such a view cannot be reconciled with the introduction of anything metaphysical. Creationism is accordingly ruled out because it is a metaphysical, not a scientific, concept. Even Traducianism is ruled out for the same reason since, although once introduced it could conceivably fit into an evolutionary pattern, creation *ex nihilo* is still required to start the process off.

Thus mind or will or consciousness is viewed as a misinterpretation of the evidence. These things do not exist in their own right but are merely a spin-off from electrochemical processes. *Behaviourism* is really the only acceptable form of psychology for the strictly scientific mind. J. B. Watson, the founder of the Behaviourist School, long ago said that the time had come to eliminate consciousness altogether from the vocabulary of the psychologist!

The present trend is thus to convert the responding person into a mere reacting thing, and so to categorize all behaviour as reflex. Man has to all intents and purposes been annihilated. Soul or spirit simply do not exist as separate realities.

Bruno Bettelheim has recently published a book dealing

with the concept of *soul* in Freud's psychology.* He shows that while Freud unashamedly employed the word for *soul* (*die seele*) throughout his works, his American translators avoided the word soul like a plague, using circumlocutions such as "mental personality". Bettelheim suggests the reason for this was that the psychologists felt Freud's ideas would be more readily accepted within the scientific community if the concept of soul was entirely eliminated!

We are, in their view, merely a piece of biological machinery. If the reactions of the machinery of the body *can* be called 'soul', then soul is nothing more than an epiphenomenon, a secondary effect which has no existence in its own right and exerts no influence on the object which gives rise to it, any more than the babbling of the brook influences the movement of the water which babbles . . . There is no ghost in the machine.

The strictly logical evolutionist must either adopt *panpsychism*, the doctrine that consciousness was a characteristic of matter from the start so that there is no such thing as unconscious or inanimate matter; or he must admit it suddenly emerged out of the blue without antecedents, thereby creating a discontinuity in nature.† Neither position is felt to be tenable, so the "horns" of this dilemma are simply swept under the carpet and no longer discussed.

It is interesting to note that the idea that all matter must be considered as already animated was favoured by Zeno who wrote in his work *Concerning the Nature of the Gods*, "Nothing that is without a soul and reason can generate of itself anything endowed with life and reason; the world however generates beings with soul and reason; therefore the world is itself living and possessed of mind".

---

* Bettelheim, Bruno, *Freud and Man's Soul*, Knopf, N.Y., 1983.

† A discussion of this question in the light of modern research will be found in the author's *The Mysterious Matter of Mind*, written for *Probe Ministries*, Dallas, and published by Zondervan, Grand Rapids, 1980.

This was written nearly 400 years BC! In man's way of reasoning little has changed since then. The July issue of *Science Digest* (1981) has an article entitled "Is the Cosmos Alive?" Both are really inspired by the same problem: Whence came man's soul? Where did *consciousness* come from? The evolutionary concept of gradualism without any discontinuities seems to demand that consciousness is a property of all matter and has co-existed with it from the very first.

But this alternative, that matter generated consciousness, really does not solve the problem. It merely shifts it one step further back, since matter itself has to be accounted for, and the *eternity* of matter is no more conceivable than that it had a beginning at some point in time. Carl von Weiszacker tried to resolve the difficulty by arguing that matter and spirit are really one and the same thing. The Christian who believes that God created both matter and spirit really has no problem with this concept, because matter simply becomes an epiphenomenon of spirit rather than the reverse! As Hebrews 11:3 puts it, "By faith we understand that the . . . . things which are seen were not made of things which do appear". For the fact is that both visible *and* invisible things are made by God who Himself is invisible (Col.1:15, 16).

So we have these four alternatives with regard to the origin of the soul in each individual. First; it is a reincarnation of a soul already in existence. The soul has thus passed from life to life in a process of purification and ultimate absorption into some state of rest which relieves it of further embodiment. This is really not an account of its origin but only of its subsequent history. Secondly; that each soul is a separate creation of God, infused into the procreated body at some appropriate instant. Thirdly; that it is generated by the parents as the body is generated, each parent contributing their soul-stuff. Lastly, that it is a kind of "static noise" emanating from the electro-chemical activities of the brain and has therefore no separate existence.

"Reincarnation" solves no problems of origin but only suggests a destiny. "Creation" creates problems for anyone who cannot accept the concept of discontinuity but demands an unbroken "great chain of being". "Traducianism" still leaves the matter of the *origin* of soul-stuff unanswered.

Meanwhile "Evolution" has to deny that my *self* has any real existence. I may deny this for others: to deny it for oneself seems to me to be virtually impossible in a normal mind. Descartes' "proof" is pretty compelling. Translated freely, he argued thus: "If I doubt, then I must exist just to entertain such a doubt"! Therefore my personal existence cannot be called in question.

The best that Julian Huxley could do was to admit frankly that the origin of mindedness is a "glorious paradox"; and since it can hardly be denied, it serves only to prove how marvellous the evolutionary process really is because it can do such incredible things! One cannot rationally come to grips with circular reasoning such as this . . . .

If the law of parsimony is allowed to govern our thinking in the matter, we ought perhaps to admit that direct creation is the simplest explanation and has greater explicit support from the New Testament than any other alternative.

The evolutionary concept leaves us on the horns of a dilemma. Either soul-stuff belongs to all matter or it appeared suddenly and thus forms a discontinuity in the scheme of things. While panpsychism seems absurd, any discontinuity is equally unacceptable. One has to make a choice between two unallowables!

Chapter 9

THE GHOST IN THE MACHINE

Evolution is essentially a philosophy of materialism. No self-respecting evolutionist will countenance the view that evolution provides only a partial answer to the existence of a living world. Matter, life, consciousness, and self-consciousness in an ascending scale form part of a great chain in which there are no discontinuities, nothing that cannot be quantified and ultimately understood in electrochemical terms.

The chain is deterministically linked and must be preserved without the introduction of any mechanism or any source of energy that is not absolutely part of the system. The Universe is a uni-verse, not a multi-verse, and one set of laws governs all that happens within it. Divine interventions are unallowable.

Thus if there is such a thing as soul or spirit or will or self-conscious mind, it is not another order of reality but a direct outcome or spin-off of matter that has reached a certain stage of complexity. Ernst Haeckel (1834-1919) in an address before the German Association (of Science), assured his audience that once the chemical constitutents of a cell (carbon, hydrogen, nitrogen, and sulphur) are correctly assembled, they "produce the soul and the body of the animated world, and suitably nursed, become man . . . . With this single argument the mystery of the universe is explained, the Deity is annulled and a new era of infinite knowledge is ushered in"!

Today, those evolutionists who are strictly logical in their thinking ought not to find anything to quarrel with in this statement. Although they admit that there are still many unanswered questions, they continue to have confidence in the thesis. For many generations evolutionary biologists and physiologists have interpreted their research findings in strict compliance with this philosophy. Thus, when in recent years a series of discoveries have been made that challenge the view that man is effectively a machine by "rumours of a ghost inside", all were surprised – including the finders themselves.

One of the most impressive lines of evidence of this "ghost" has come from the work of Wilder Penfield who died recently after many years of highly successful treatment of epileptic patients at the Montreal Neurological Institute. During these years, Penfield brought relief to well over a thousand subjects by a brain operation which was daring indeed both in its conception and its execution. What he found was as much a surprise to himself as it was to his colleagues.

Now, to keep the record clear, it is appropriate to note the very important distinction that is being assumed between mind and brain throughout the rest of this chapter. The mind is the *thinking*, the *will*, the *conscious attention, something non-physical*, the *self* which scrutinizes the situation or contemplates its meaning; the brain by contrast is the *organ* which the mind uses, a physical object located in the skull and composed of billions of nerve cells each of which may have ten thousand connecting fibres to neighbour cells – an incredibly complex structure.

Penfield began his life work in the belief that the brain accounts for the mind as the brook accounts for the babbling noise it makes. Stop the flow of water and the babbling ceases; destroy the brain and the mind is destroyed. Mind and babbling are epiphenomena – not realities with independent existence.

Before describing this work, let me quote something he wrote in retrospect after his retirement: "Throughout my own scientific career, I, like other scientists, struggled to prove that brain accounts for the mind . . . . Now, perhaps, the time has come when we may profitably consider the evidence as it stands, and ask the question: *Do brain-mechanisms account for the mind?* Can the mind be explained by what is now known about the brain? If not, which is the more reasonable of the two hypotheses: that man's being is based on *one* element, or on *two*?"

Very briefly, the technique which Penfield used was to lay bare a segment of the cortical surface of the brain by removal of a portion of the skull. The motor area thus exposed is known to be related to the involuntary movements of epileptic subjects. The patient was lightly anaesthetized locally for the opening up of the skull and felt little or no pain, but was kept fully conscious because his or her ability to communicate with the surgeon was essential to the success of the operation.

The surgeon was thus enabled to probe the surface of the brain in search of the damaged area. The probe itself was composed of a single electrode, using a 60 cycle 2 volt current. Contact with the brain surface caused no sensation of pain whatever, but when the damaged area is thus stimulated, the patient at once becomes aware that an epileptic fit is pending. By this means the offending tissue could be located and, hopefully, corrective surgery applied.

Rapport between patient and surgeon had to be at all times extremely close, and Penfield himself inspired enormous confidence. But what he discovered unexpectedly was that the electrode stimulation of very specific areas brought an experience of recall for the patient which was so vivid that the subject frequently suspected Penfield of using tape recordings. The recall involved sound, coloured vision, and even odours. Any one area could be contacted again and again and the recalled events would be perfectly replayed in great detail in the patient's consciousness. On one occasion the electrode was applied to the same spot 60 times and the recall experience was on each occasion precisely the same.

What was so surprising was that the recall in no way at all interfered with the patient's full awareness of all that was taking place in the operating room, including conversations among the staff and even street noises outside. The subject was experiencing two kinds of consciousness: one from some long past situation, and the other from the circumstances surrounding the operation. It was therefore quite possible for the subject to live in two worlds and to discuss in great detail the recalled events with those who occupied the operating theatre. It was also quite possible for the patient, wherever the recall involved music, to both whistle or hum the piece being played: and often even to identify it.

One day, Penfield had an experience which opened his eyes in a new way to the existence of a ghost in the machinery. Here is how he described the event.

> When the neurosurgeon applies an electrode to the motor area of the patient's cerebral cortex causing the opposite hand to move, and when he asks the patient why he moved his hand, the response is: 'I didn't do it. You made me do it . . . ' It may be said that the patient thinks of himself as having an existence separate from his body.
>
> Once when I warned a patient of my intention to stimulate the motor area of the cortex, and challenged him to keep his hand from moving when the electrode was applied, he seized it with the other hand and struggled to hold it still.
>
> Thus one hand, under the control of the right hemis-

> phere driven by an electrode, and the other hand which he controlled through the left hemisphere, were caused to struggle against each other. Behind the 'brain action' of one hemisphere was the patient's mind. Behind the action of the other hemisphere was the electrode.

Penfield was driven to conclude, therefore, that the brain was not causing the mind but was the servant of it and its instrument for willed action. This chance finding was repeated many times and has since been experimentally demonstrated beyond a shadow of doubt. The brain controls the action but the electrode controls the brain. In all willed action, precisely the opposite is true: the mind controls the brain and uses it to effect its purposes upon the material world. To the extent that the brain is a physical organ whose development has resulted from a combination of genetic endowment and the accidents of experience, it must be said, as Viktor Frankl put it, that while the brain *conditions* the mind, it does not give rise to it.

A second striking proof of precedence of will over matter, of mind over brain, has now been provided by some elegant experiments by H. H. Kornhuber.

To describe his findings with scientific precision would be to snow the reader, and lose him! But very simply, here is what he discovered. Whenever an action of any kind is willed, measurable electrical potentials are generated in the motor area of the cortex that controls the action. These potentials are observed *prior* to the action that is willed but only after the "willing". Thus between the conscious exercise of will and the activity that results there is a clearly measurable delay of sometimes up to several seconds duration.

In this brief but significant interval, there is a flurry of electrical impulses over a wide area that gradually narrows down and concentrates the signal to bring about the precise movement willed. The delay between willing and activity is quite measurable, and the nature of the will and the resulting activity correspond.

In commenting on this discovery, the neurophysiologist Sir John Eccles remarks, "it is rather like the sergeant saying 'Company . . . .' before giving the specific command which is to follow. It seems to warn that the will is about to act on the mechanism".

Eccles concludes, "Thus we can regard these experiments as providing a convincing demonstration that voluntary movements can be freely initiated independently of any determining influences within the neuronal machinery of the brain itself". In short there really *is* a ghost in the machine, capable of giving orders to the machinery and able to use the machinery for its own purposes.

The problem that still remains, of course, is *how* the mind or will (or, in our context, the spirit) actually makes contact with the brain. I can normally move my hand at will, but I cannot move the hands of my clock on the wall. Clearly there is no connecting link with the clock hand over which to mediate the message. But if my will is a non-physical spirit, where is the actual connection with my own brain that makes it move my own hand? The process is just as elusive. It has been suggested that if we could really resolve this problem, we might have a better idea of how God acts upon the physical world, of how faith could move mountains; even of how the Lord instantly stilled the violent storm on the Sea of Galilee.*

Descartes recognized this interaction of spirit and body, and also that it works both ways. Today we speak of these two kinds of interaction as psychosomatic and somatopsychic, i.e., spirit/body and body/spirit interactions depend-

---

*** Perhaps the Lord Himself can take over the autonomy of the human mind to use that mind's brain to effect a desired end: for example, in the writing of Scripture, or in giving skill to the hands — as He did to Bezaleel for the embellishment of the Tabernacle (Exod.36:1). And perhaps there is more truth than we realize in the saying that the devil finds work for idle hands to do.**

ing upon which takes the initiative. But like Descartes who finally abandoned the search for the "contact point", we too have come little or no nearer to a solution.

However, Descartes did uncover a phenomenon that we all recognize readily enough when it is pointed out to us in the right way, a phenomenon which clearly demonstrates (as Penfield and Kornhuber have shown) that there is someone there in the machinery making it work for us.

What we have in mind is the ability we all possess to see the objective world stereoscopically, that is to say, in depth, in three dimensions and not just in two as an ordinary photograph presents it. We can perceive distance between objects and we can perceive the roundness of things, so that they are neither all at the same distance from us nor are they flat.

We are able to do this because we have two eyes spaced on average 650 mm. apart. By using a camera with two lenses similarly spaced so that two separate photographs of a scene can be taken at the same instant from these two different positions, and by mounting these two photographs appropriately so that each eye sees only one picture, we suddenly find ourselves back in the three-dimensional world.

I have, and use, just such a camera and never cease to be thrilled by the steroscopic effect. One wonders why these cameras are not more widely used, except that one needs a special viewer.

Now the interesting thing is that when one first looks into the viewer, one is met by two *different* pictures. They are very little different, though they are momentarily irreconcilable. The difference between the two pictures is clearly revealed if we superimpose the one over the other and project them together on to a screen. The picture on the screen will be fuzzy and blurred, rather like a photograph taken by a camera that has been moved slightly at a critical moment. *No amount of focussing of the projector will resolve the blurred or fuzzy picture.*

I have emphasized this because it is of crucial importance. The fact is that the two pictures cannot be reconciled in any way by mere overlapping projection onto a screen, *nor even by projecting the two pictures through the lenses of the two eyes on to the two retinas*. Descartes presumed that the signals from the two retinas which are screening two irreconcilable pictures reach a single place in the brain where they coincide and fuse themselves into a single picture in three-dimensional form. He was never able to trace the nerves, or he would have discovered that these two signals do not fuse as a single picture in the brain but remain as independent signals in the visual cortex, one in the left hemisphere and one in the right hemisphere.

How, then, do the two conflicting scenes come to be viewed as one? The conclusion is that the *MIND* and not the *BRAIN* fuses and synthesizes them. So far it has been impossible to say how this is done.

The fact is that when anyone for the first time looks through my viewer, he or she will find that the two views are *not* reconciled at once. But then quite suddenly, in a way which is very hard to define but is a very real experience, they are reconciled and two contradictory pictures in the flat are instantly experienced as one in the round. The 'shift' is remarkable and is very sudden. The two scenes converge unexpectedly and, *voilá,* the picture emerges in three dimensions! It is an interesting experience and a wonderful demonstration of "mind" at work.

Moreover, from the moment the viewer has achieved *one* three dimensional resolution, thereafter one can go through several hundred pictures, and there is no such experience of delay and sudden resolution as occurred in the first viewing. Indeed, one develops a kind of 'pre-set' facilitation that allows reconciliation to proceed much more rapidly, even long after having laid aside the viewer. Days and months can intervene, but the skill to obtain immediate stereovision is not lost. It would appear that the mind has learned the trick of it.

It is not the two eyes which look out upon the distant scene that give us a view of things in the round: it is the *fusing* power, the reconciliation of two quite discrete and

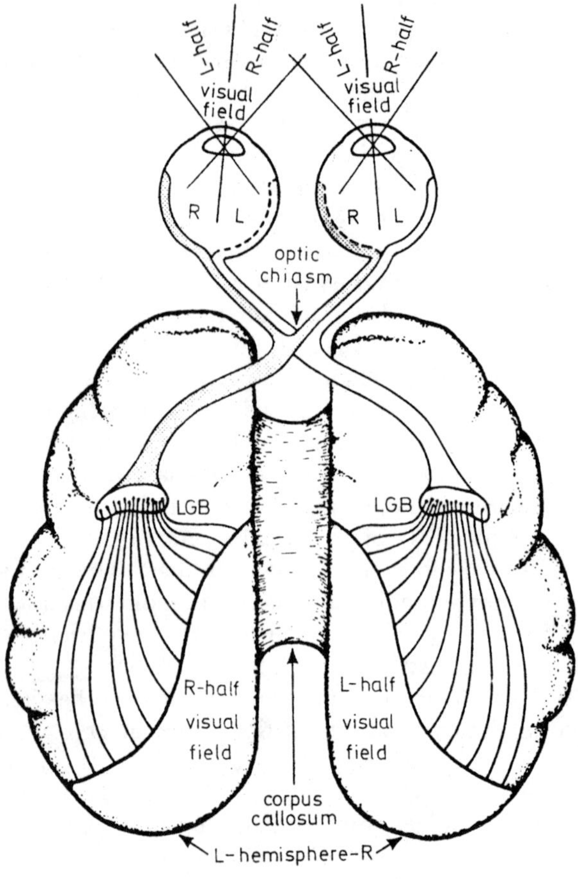

Diagram showing how the pathways of the separate signals from the two eyes lead to two separate areas of the brain.

subtly conflicting pictures that is done by the *mind.* The brain provides the materials in coded form: the mind or soul or spirit uses these discrete materials to form a view of reality that corresponds to reality in depth. Without this *fusion*, all would be *con*-fusion. When the two signals are presented incorrectly due to a fault in focussing power of either one of the eyes which may have occurred as a result of a breakdown of the machinery, it is the mind or soul or spirit which compensates by ignoring one of the two inputs and thus resolves the irreconcilable conflict. It is true that when this happens we lose depth perception but at least the mind "makes sense" out of what the *eyes* are signaling confusingly. The mind can only fuse what the brain is presenting in the correct relationship.

Were it not for this power of the mind to do what neither the camera nor the eyes themselves can do, we would have to cover one eye to eliminate the contradictions in the two pictures! But *without* stereovision we could not perceive the depth or thickness of things, nor reach out with perfect confidence and grasp them, nor even thread a needle except with some difficulty and by constant trial and error. To drive a car, we should have to gauge our distance from the car ahead by its changing size, and the picture of the world we have would be flat. We would get used to it. People with only one eye do. But many kinds of "comprehension" of reality in the physical world would be far more difficult, if not impossible. It is doubtful if the soldering of an extremely small electronic circuit *manually* could be done. In a thousand ways there would be confusion until we had learned in each situation and at that particlar moment how to relate to space and distance. By it, we know where we are within the framework of things.

The mind does for us what the eyes cannot do, what the stereocamera cannot do, what the stereomicroscope cannot do, what the stereoprojector cannot do – in short, what no *machine* can do. It does this by fusing two flat pictures that conflict, into one picture which provides the added dimension of depth. Without the mind to manage the input, there is no creation of order, no meaningful fusion. The machine itself no more does the fusing than the computer "adds". There is no *adding* going on in the computer. It is just a machine re-routing signals according to design in such a way as to give us a signal meaningful only to the *mind*. The computer doesn't *know* what it is doing.

Now it is obvious that many animals below man have stereoscopic vision, though not over quite as wide a range in some cases. It must be assumed, therefore, that in these animals, as in man, there is a ghost in the machinery integrating the two discrete signals reaching the brain. Does this then mean that animals also have a body/spirit constitution, so that they too are *souls*?

The answer appears to be, Yes. Indeed, the Hebrew word for soul (*nephesh*) is frequently used of animals and was actually applied to animals (in Gen.1:20, 21, 24 and 30) before it was applied to man! Is there no difference, then, between man and the animals in this regard? There certainly is, for although the fate of the bodies of both men and animals is to return to the dust, the fate of the spirit is quite different. Ecclesiastes 3:21 tells us that the spirit of man goeth upwards (i.e., returns to God who gave it, Eccl.12:7) but the spirit of the beast goes downwards to the earth to share the fate of its body. In other words, the destiny of the two is diametrically opposed.

We have been using a number of terms such as *mind, mindedness, will, soul, spirit,* etc., indiscriminately. It is unfortunate that there are so many terms to describe the activities of the ghost in the machine. All of these terms are appropriate and each one of them is essentially a spiritual faculty rather than a physical one. I want to say a word about what is meant by *consciousness* and the even more difficult-to-define phenomenon, *self*-consciousness.

Children are occasionally born without a brain. Yet they react to sounds and odours and to physical contact.

In spite of this reactivity, there is no evidence of actual consciousness and, of course, there is no brain to mediate it. Thus, in effect, there is "no one there". They are alive but act like a person in a deep, deep coma, except that they never will and never can come out of it. They have no conscious contact with outside reality because they have no cortex. This situation points up the fact that a surprising level of reactivity is possible in the total absence of consciousness.

If animals have consciousness, then one must ask whether there are levels of consciousness corresponding to the level of development of their central nervous system. Plants, of course, "sense" the environment by various means without a central nervous system and therefore presumably cannot be said to have consciousness. Animals, on the other hand, even the simplest of them like the amoebae, have consciousness at least of *other bodies* around them: but man is intensely self-conscious, i.e., conscious uniquely of his *OWN* body, both in times of health and in sickness.

The last point is important. Animals seem obviously conscious of their own bodies in times of sickness or when wounded, but man is conscious of his own body most of his waking life. When a baby first discovers its hands and feet are part of itself, the first glimmerings of self-consciousness have probably been born. It is to be noted, therefore, that in man the body introduces itself early in life and engenders a new *kind* of consciousness, consciousness not merely of other bodies but of one's own. We see our own bodies as something possessed, something we can stand apart from and consider objectively.

In this matter of personal body identity, I think it is highly significant that man is the only creature who cares for the bodies of his dead – including his own! No animal is known to take anything more than a momentary and passing interest in a dead body (except as food of course), not even the dead body of its own mate. No attempt is made to protect the body from predators once it is dead.

But even from the earliest times, as is known by the presence of flowers in the grave, man has buried his dead quite deliberately. Not infrequently the dead are placed in a foetal position, a fact which some suggest is witness to a hope of re-birth or reincarnation.

Theodosius Dobzhansky noted "that all people everywhere take care of their dead in some fashion, while no animal does anything of the sort". And it is a significant fact of human behaviour that a dead body continues to be protected by custom or by law from injury. Indeed, we seem to have stricter rules about the mutilaton of the dead than we do of the living. Man's consciousness of his own body or of the bodies of his fellow men seems to be very different from mere animal consciousness.

Sir John C. Eccles (Nobel Laureate for his work in neurophysiology) in 1977 co-authored with Sir Karl R. Popper a fascinating book supporting the reality of the soul or spirit and thus the dual nature of the human constitution.

In the process of writing the book, they had planned to title it *The Self and The Brain*. But subsequently, they both became so convinced by the consideration of the steadily accumulating experimental evidence of a managing spirit or mind within the machinery, that they changed the title. The change was a very small one from the typesetter's point of view, but it was a highly significant one in its implications. The new title became *The Self and* ITS *Brain*.

Eccles concludes that the self-conscious mind is not simply engaged in passively observing the signals from its body but in actively sorting them out. Continuously displayed before it is a whole complex of neural inputs from the eyes, the ears, the nose, and – be it noted – the skin (which is the largest organ of the body). It selects from this chain of signals in the brain whatever is of interest to it, blending the result from different areas of the cortex. The body supplies a rich input, and the mind deliberately disregards some and attends to others.

In this way the self-conscious mind achieves a unity of experience which becomes entirely *personal*. The brain is its own brain, a personalized computer that it uses. According to this hypothesis, the prime role in this process is played by the self-conscious mind, exercising both its selective and its integrative abilities.

They were impressed with the bond which exists between spirit and body, a bond often commented upon by theologians in the past from Thomas Aquinas (1224-1274) to James Orr (1844-1913), and right down to a number of present writers. All have sensed the closeness of this bond which arises out of or generates in man a strong sense of personal identity, far more profound than is to be observed in any animal. It is a form of consciousness that is related entirely to the *spirit's* awareness of its own *body*, which is now acknowledged by some of the best modern students of animal life as being unique to man.

James Orr made much of this bond and attributed to it the abhorrence of physical death which seems to have characterized man's thinking throughout history. The spirit in man, though burdened by the body that spoils so many of his highest aspirations because of its demands, nevertheless is so strongly attached to it that it fears the rupture of death throughout life. The most dramatic and most perceptive definition of death, and the truest theologically considered, is still "the separation of the spirit from the body". This is what death is.

We long to be freed because it is a house in ruins, but we tremble at the prospect of this separation as one might tremble at the loss of a companion of a lifetime, and that is what the body has been.

Mind or spirit is the seat of authority. The brain is an instrument which serves it. The whole is integrated, unified, ordered by the mind. This non-physical reality is master of the meaningful operation of the body which is, ideally, its servant, but due to the Fall may and does become its master all too often.

This struggle is graphically described by Paul in chapter 7 of Romans. The Fall had consequences as fatal to the body as to the spirit. We see interaction for both good and ill exemplified throughout Scripture.

The Lord remarked sadly upon it in Matthew 26:41, "The spirit is willing but the flesh is weak", as He came to his companions in his terrible hour of anticipation and found them, alas, asleep. Paul bewailed it when he cried out, "O wretched man that I am! Who shall deliver me from this body of death" – while at the same time saying that he really had no desire at all to be unclothed, i.e., disembodied; though meanwhile he groaned in the body he had (2 Cor.5:4). Here was an unwanted interaction.

But the Bible speaks of co-operative interaction as well. In a passage seldom noted (Mark 9:29), after the disciples had come to the Lord in surprise that their newly delegated power to heal sickness had failed them, Jesus said, "This kind cometh forth by nothing but by prayer and fasting". By *prayer* which is a discipline of the spirit, and by *fasting* which is a discipline of the body. There are some battles with Satan that require *the whole man*, body and spirit – despite the fact that the battle is a spiritual one.

Man has a brain that is clearly a kind of computer. This is what man HAS, but it is not what man IS. Man is truly a spiritual creature but he is an incarnate, embodied creature, unique among all other creatures because of the uniqueness of the origin and destiny of both his body and his spirit.

It is remarkable that a man of Eccles' stature and experience who nevertheless does not share our Christian convictions, should arrive at this same conclusion, and in doing so should reject the evolutionary origin of the soul. He did so in spite of the fact that his co-author, Karl Popper, accepted it without equivocation.

In another of his recent works, *Facing Reality*, Eccles observes that statements about "the progressive emergence of conscious mind during evolution are not supported by any scientific evidence, but are merely statements made within the framework of a faith that evolutionary theory, as it now is, will at least in principle explain fully the origin and development of all living forms including ourselves".

This "faith" is being steadily eroded by experimental evidence. There IS a ghost in the machine.

Chapter 10

A HOUSE IN RUINS

**I know that in me, that is in my flesh, dwelleth no good thing.
For we that are in this house do groan, being burdened:
not that we would be unclothed.**

Rom.7:18; 2 Cor.5:4

C. S. Lewis wrote in his book, *The Problem of Pain*: "(The Fall of Man) was not, I conceive, comparable to mere deterioration as it may now occur in a human individual; it was a loss of status as a *species*. What man lost by the Fall was his original specific nature . . . .

"This condition was transmitted by heredity to all later generations, for it was not simply what biologists call an acquired variation; it was the emergence of a new kind of man – a new species, never made by God, had sinned itself into existence. The change which man had undergone was not parallel to the development of a new organ or a new habit; it was a radical alteration of his constitution, a disturbance of the relation between his component parts, and an internal perversion of one of them.

"Our present condition, then, is explained by the fact that we are members of a spoiled species."

Since we are as much a body with a spirit as we are a spirit with a body, it is clear that a "spoiled" body makes a spoiled person. Our nature in its entirely, body and spirit, has been fatally injured. We are, in fact, no longer MAN at all, no longer MAN as God intended.

When the Lord God called out in the Garden of Eden, "Adam, where are you?", it was, I suggest, not the man hiding in the bushes that He was looking for. He knew where that individual was. What had been lost from the web of life which He had just finished creating was the master species, the appointed agent of management for that web of life. Mankind was, in short, the first of a whole catalogue of species which would subsequently be "endangered". The extinction of species, so common to history since, began with the Fall of the species, Man.

The Lord God had just finished creating Man as the one species that bore his own image. It was *that* kind of "man" that had now disappeared, converted by sin into a creature quite unlike the original – without innocence, without immortality, without the sure instincts of all other creatures, and above all, having lost the image he had been endowed with.

Adam's children were born in his image, no longer in God's image. Genesis 5:3 makes this only too clear by stating the two positions very deliberately and in juxtaposition. God had created man *in His own image* (verse 1): fallen Adam now procreated men in *his* own image (verse 3).

By the introduction of a deadly poison into his system after eating the forbidden fruit, he had entailed to all his naturally born descendants a fatally flawed constitution, both physically considered and spiritually. And the process has been at work generation after generation, steadily deteriorating man's vitality from a life span of nearly a thousand years to 120, and for the vast majority of his descendants considerably less than that – by David's time a mere three-score and ten (Psa.90:10).

The world was not to see another truly Adamic body as God had created it, for four thousand years: not until it re-appeared in the Person of the Lord Jesus Christ, in a form so superb, so magnificent, so beautiful, that it proved a perfect vehicle for the expression of God Himself.

Now and then we see a 'splendid specimen' of manhood or womanhood, a splendour which survives for a few years. But hiddenly it is already dying. All the time a secret rot is eating away at it. When finally, with the departure of the spirit, death blows out the candle a process of decay that has been proceeding throughout life suddenly gallops away, and disintegration takes place with appalling speed – unless the undertaker quickly intervenes to slow it up.

This decay is not something that has just begun. Living, we are dying – from the moment of birth. The transformation of an immortal creature into a mortal one has affected every cell, every organ, every fibre of the body, and leads inevitably to the transformation of what has had beauty in its day into what now tends only towards ugliness or senility.

It is only the extraordinary refinement of this machinery, with its ability to heal wounds and to correct its DNA transcription errors, that preserves it as well as it does.

Now, is there any evidence that the corruption of the body does truly 'insult' the soul or spirit in any very vital way? Indeed there is. And the fact was made remarkably manifest when men and women by the hundreds of thousands were subjected to the desecration of their bodies in the Nazi concentration camps of World War II.

Many studies of these camps and their effect upon people have been undertaken in recent years. One of these, by Terrence Des Pres, is titled *The Survivor*. It is a study of what factors enabled some few to survive while the vast majority were simply overwhelmed and died in the camps.

One of the remarkable things that Pres observed after interviewing or corresponding with a great number of these survivors, was the importance of care of the body, *of cleanliness*, even if it was pitifully little they could do – usually amounting to nothing more than a token washing by dipping a finger in their cold ersatz coffee and touching the cheeks or the forehead.

One survivor whom he interviewed observed, "I began to look around me and saw the beginning of the end for any woman who might have had the opportunity to wash and had not done so". These people lived in filth of unimaginable dimensions – excrement, vomit, running wounds, no form of uncleanness was lacking in that environment. In some camps they stood ankle-deep in it all day. Even the birds soon ceased to fly over the camps because of the stench. And yet, the tiniest, wee-est token of attention to the cleanliness of the body was often enough to keep the spirit alive. When that token was abandoned, the individual was already as good as dead.

Another survivor said, "Many inmates ceased to wash. This was the first step to the grave. It was almost an iron law. Those who failed to 'wash' every day soon died . . . . an infallible symptom."

Des Pres is a most perceptive writer, He comments on this: "If spiritual resilience declines, so does physical endurance. If the body sickens, the spirit begins to lose its grip. There is a strange circularity about existence in extremity. Survivors preserve their dignity in order 'not to begin to die': they care for the *body* as a matter of '*moral* survival'." One could scarcely ask for a more striking illustration of the interaction and interdependence of body on spirit and spirit on body.

Pres speaks of the defilement of the body reaching such proportions as to produce what he aptly terms "spiritual concussion". Some were forced to eat excrement and if they refused, had their heads held under until they complied. When they were allowed to lift their heads, they literally went insane. One survivor spoke of immersion in human excrement as "the nadir of his passage through extremity. No worse assault on a man's moral being seems possible".

Women who have in our day and world been raped – sometimes by more than one attacker – have had such a devastating sense of defilement as to attempt suicide. Those who are acquainted with the story of Lawrence of Arabia may recall that he, too, felt like committing suicide after being defiled by a despicable minor Egyptian official.

I think Pres is quite right to stress the fact that a feeling of defilement underlies the concept of guilt and that washing of the body underlies the concept of spiritual purification. The association between moral cleansing of guilt and physical purification from defilement, seems to be reflected by the many occasions upon which ritual washing is prescribed in the Bible for those engaged in the service of the Lord. See, for instance, Exod.30:17-21; Lev.8:6, and in the New Testament Acts 22:16; 1 Cor.6:11; Heb.10:22. And as having an obvious bearing, see also Ezek.36:25; Zech.13:1, and John 13:10.

It may be argued that cleansing of the *body* without cleansing the *spirit* is ineffective, except for social reasons. The observation is clearly correct. But by the same token, it may well be that cleansing the spirit without cleansing the body would seem to be equally ineffective. The highly spiritual individual who doesn't care for the cleanliness of his person can only be half-highly spiritual!

But the normal reaction to this kind of antithetical statement would probably be, "Well, the body doesn't matter nearly so much as the soul, does it?" Hebrews 10:22 bears on this matter: "Let us draw near with a true heart in the full assurance of faith, having our hearts sprinkled from an evil conscience (i.e., our *spirits washed in the blood* of the Lamb) and our *bodies washed with pure water*." This two-fold cleanliness of both spirit and body seems to be required for a fully mature faith, for a whole not a half faith.

It is an odd situation, this ambivalence we have about the value of the body. Here we have a tumbled-down house for the spirit, which the spirit is nevertheless deeply attached to – so deeply that it faces separation with grave concern. And this grave concern is just as likely to be shared throughout most of life even by those who have every hope of a far more excellent house reserved in heaven.

There is no doubt that the Bible paints a rather awful picture of the state of ruin this house is in. Paul is particularly strong on this point.

In Romans 6:6 he refers to it as a "body of sin".

In Romans 6:12 he speaks of it as "mortal". This might be harmless enough if mortality was *natural* to it. But it isn't at all natural to it. It was designed for immortality, so that its mortal-ness is precisely what it was *not* designed for. It is in an un-natural state, a condition of fatal disease tantamount to a kind of leprosy that in death places the body, according to Scripture, in the same "untouchable" position as the leper is in during life. Neither the leper nor the dead body can be touched by the living without suffering defilement (Cf. Lev.13:44-46 and Num.19:11-20).

Hiddenly our living body is as inwardly diseased as a leper's body is outwardly so. And this is because it has been unnaturally mortalized and is, in fact, already as good as dead.

Indeed, Romans 7:24 refers to the body as a "body of death". This is an expression which is Hebrew *in content* but Greek *in construction:* it would have been better rendered simply "a dead body".

This form of transposition of words is common in Hebrew. It appears, for example, in Psalm 48:1. "the mountain of his holiness", which is better rendered "his holy mountain". So in Psalm 47:8 we should read "his holy throne" and in Job 30:12 "their destructive ways". So also in the New Testament we have from Paul in Ephesians 6:10, "the power of his might", i.e., "his mighty power"; in Colossians 1:22, "in the body of his flesh", meaning "in his fleshly body".

When in Romans 7:24 Paul refers to this dead body which he longs to be delivered from, he may have had in mind a situation which was not uncommon in the Roman world. One particularly awful form of punishment for a convicted murderer was to have the victim's body chained to him in such a way that he was forced to carry it or drag it wherever he went until in its final state of decay it literally fell to pieces . . . . If this is true, it is no wonder that Paul, in Philippians 3:21, should speak of the body as something that is vile.

The reader may be perfectly justified in protesting that Paul's body was overtly diseased in some way which scholars have not yet agreed upon, and that therefore his thinking was highly coloured by personal experience. But the important thing is to recognize that when man dies, he dies an unnatural death, a death which he has been dying all his life. For many this process is delayed in such a way as to conceal the fact of decay and almost to hold out a promise of immortality. But as soon as the spirit departs, the illusion is destroyed. The disintegration of the body is rapid indeed. And it is doubtful if man finds anything quite as distressing to look upon as a decomposing human body. It is a terribly disturbing sight for man – though apparently animals are almost if not totally indifferent to it.

Only in the presence of such decay is the distance made plain between the body of fallen Adam and the body of unfallen Adam as revealed in that of our Lord Jesus Christ. Such was *that* body that even after burial for three days it saw no corruption. And it saw no corruption because it was never subject to death in the first place.

Our dying is always in some measure "an act of violence", the tearing apart of two elements – the spirit and the body – which were never intended to be parted. But it was essential in God's economy of things that this body should be destroyed since it has become a partner in our fallen nature and only by dissolution and resurrection in an entirely new form could it be rid of its infection. The spirit of man is newly re-created (2Cor.5:17), not merely reformed: the body of man cannot be merely reformed either. We dwell in a house in ruins which, after it has fallen to pieces in the grave, will not simply be put together again. It will not be re-formed out of the old substance, but transformed into a house as glorious as that which is Christ's in his resurrection (Phil.3:21).

Thomas Boston in 1720 put the matter very beautifully when he wrote: "There is a vileness in the body which, as to the saints, will never be removed, until it be melted down in the grave, and cast in a new form at the resurrection, to come forth a spiritual body."

Man lives in a house in ruins, in a body which does not represent a mere miscueing of evolutionary processes but a body which was designed for, and in two individuals displayed, a potential for physical glory and immortality. Both individuals were called Adam.

Evolution gives us no clue as to how *they* appeared on the scene, for the first was formed by divine creative intervention and the second miraculously conceived and virgin born.

Chapter 11

A HOUSE OF GLORY

The house that is to be builded for the Lord
must be exceeding magnifical

1 Chron.22:5

Anselm of Canterbury in England (c. 1033-1109) wrote a very remarkable book which he titled (in Latin), *Cur Deus Homo*, i.e., "Why God Became Man".

At one point in a simulated conversation with a friend he discusses the various ways in which the Saviour might have become man. Here is what he said.

> In four ways God can create a man; namely, either of a man and a woman in the common way; or neither of a man or of a woman, as He created Adam; or of a man but not of a woman, as He created Eve; or of a woman without a man, which thus far He had never done.
>
> Wherefore, in order to show that this last mode is also within his power, and was reserved for this very purpose, what more fitting than that He should take that man whose origin we are seeking [i.e., the God-man Redeemer] from a woman without a man.

This seems to me a wonderful exercise in logical construction and effective use of the English language. Yet I am not sure that he really understood why the virgin birth was so important. But such understanding is only easier for us today because we stand on the shoulders of those who went before us.

In a volume already referred to (*The Seed of the Woman*) I have set forth at some length the relevance of some of these things we have learned since Anselm regarding how birth comes about, and in particular the bearing of these things on the birth of a *male* child without the intervention of a man. Theoretically, for genetic reasons, this is quite impossible. Yet this is how the Saviour came among us, exactly as foretold in Isaiah 7:14. "Behold a *virgin* shall conceive and bear a *SON*". It was a miracle indeed.

I have discussed in this light how and why it was both necessary and possible for the seed of the woman to be preserved against the mortalizing effects of the forbidden fruit.*

As it was in the case of Adam, so it was in the case of Eve, that the bodies of both of them were now destined to experience death. But whereas Adam's *seed* had also been mortalized, this was *not* so in the case of Eve's seed.

Because of the special design of her body, her seed was protected against mortalization, even though it was housed in a mortalized body. Furthermore, such was her constitution that she was nevertheless able, in the normal course of events, to pass on this sole fragment of her original immortality to all her female descendants. And each of them in a like manner passes on this immortal stream in every succeeding generation so long as the race continues to multiply.

It is becoming increasingly apparent now that the woman's seed, prior to its fertilization by the male seed, is the only truly immortal part of the human body left undamaged by the Fall. All other cells in the human body, male and female alike, have suffered a fatal damage *but this one priceless human heritage, the woman's seed, remains intact.*

---

* The documentation for the mechanisms involved runs to some 60 pages of fine print. This is far from being merely speculative: it is experimental fact.

Thus any child *born of a virgin* will escape the physical effects of the Fall, since the damage is passed on via the male seed. If this should ever occur *naturally*, such a child would presumably always have two characteristics. It would have the *potential* for unending physical life as possessed originally by Adam and Eve. But at the same time the child would of necessity be a female and not a male.

It is this fact which makes Isaiah 7:14 such a remarkable prophetic utterance, since clearly Isaiah could not possibly have known (except by revelation) that the birth of a *male* child from a virgin could only be by a miracle.

As to the reference to a *virgin* (and not just to a young woman, as many would like to argue), there is no doubt that *virgin* is the correct translation in English since Matthew 1:22 and 23 in quoting Isaiah confirms the fact by using a word in Greek which, for the Jews, unequivocally had this meaning. When we once recognize that the Author of both statements (Isa.7:14 and Matt.1:22, 23) is the same – namely, the Holy Spirit – then we know what that Author's intention is in the *first* instance in Isaiah by what the same Author has made quite clear in the second by employing the Greek word *parthenos** in Matthew.

Now the Old Testament is very clear that from the time of its conception a child is corrupted in body. The most obvious passage is Psalm 51:5, "Behold I was shapen in iniquity and in sin did my mother conceive me".

But Job, who almost certainly wrote long before David had penned this Psalm, is even more specific and in some ways more perceptive. He asks, "Who can bring a clean

---

*** The Jewish commentators so understood Isaiah 7:14 since they employed the same Greek word for *virgin* in the Septuagint Greek version which was produced in the second century BC for Gentile readers. For the Jews, this word signified sexual purity. It was used to describe the only kind of woman acceptable as a wife for a priest (Ezek.44:22). The idea of virginity is implicit in this word as it is found in the Septuagint (for example, Deut.22:28; Judges 19:24; 21:12; 2 Sam.13:18; etc.).**

thing out of an unclean?" And he replies to his own question. "Not one" (Job 14:4). And later in the conversation with his friends, Bildad is recorded as putting the problem that this creates even more astutely when he asks: "How, then, can man be justified with God? Or how can he be clean who is born of a woman?" (Job 25:4). Much later in time, Isaiah would state the simple fact that "we are all an unclean thing" (Isa.64:6).

So there is the problem. How is man to be redeemed, if he must find a *man* to redeem him, in view of the fact that it is impossible to find a man born of a woman who is not just as unclean as the rest of men, and therefore equally in need of redemption himself? Did Isaiah really understand the significance of his own words in chapter 7 verse 14? Did he understand why the Redeemer must be born of a virgin? Perhaps he was one of those whom Peter mentions (1 Peter 1:10-12) who pondered deeply the things they had been inspired to write yet did not fully comprehend what they had written.

What is addressed in this chapter is in essence the key to what is said in this volume. Since we were determined to keep the chapters short, it is very necessary to make sure that the point at issue in each chapter is clearly stated. In this chapter one of the best ways to accomplish this seemed to be to tabulate the structure of the argument – though this introduces a certain lack of smoothness in reading.

We are here concerned with the unique nature of the Lord's body. My plan is to demonstrate four basic points.

1. That the natural order was designed from the very beginning to accomodate the Incarnation. Part of this grand design included the mechanism of procreation by the fusion of two seeds, housed originally in a single body which was then divided into a male and a female body.
2. That the entail of Adam's disobedience was, by virgin birth through the intervention of the Holy Spirit, thus

avoided in this one instance.

3. That this led to the recovery of a truly Adamic body, i.e., an unfallen body constituted exactly as Adam's had been at his creation.
4. That the New Testament has illuminated this birth in some very striking ways and, in addition, has employed two only slightly but very significantly different words in order to make clear the distinction between that virgin-born body and the bodies of all other men.

1. The mechanism of conception prepares the way.

Many years ago Charles Augustus Briggs made the following observation: "The virgin conception of Jesus is not to be interpreted as if it were a miracle in *violation* of the laws of nature . . . . The conception of Jesus in the womb of the virgin Mary differs from all other conception of children by their mothers in that there was no human father. The place of the human father is taken by God Himself . . . in an extraordinary way unrevealed to us and without violation of the laws of maternity, impregnating the virgin Mary with holy seed".

To put the matter slightly differently, God did not *contravene* the design of the natural order when bringing the Redeemer into the world. But He put that natural order to a higher service, a service for which it was intended in the first place. He had so designed the processes of conception and birth that He could use them without doing any violence to his own creation.

He did not need to set aside nature, since there was nothing in the constitution of the human body, except in so far as it has been defiled by sin, that God is ashamed to take unto Himself and employ as a dwelling place under all the circumstances and challenges of daily life from conception to death. Undefiled by sin and indwelt by the Lord Himself, a superb human body appeared on the stage of human history, and men worshipped without shame or hesitation, and sometimes almost involuntarily, the One whose house it was. While He who possessed it, accepted the worship of men which is reserved only for what is divine, without hesitation whenever He knew it was entirely proper, and rebuked it when He knew it was not (Lu.4:41). The house that was built for the Lord was indeed "magnifical". Human embodiment in no way demeaned divinity.

2. The avoidance of the entail of Adam's disobedience.

By virgin birth was found the solution to the Old Testament questions. We can see now in the light of modern knowledge how the seed of the woman could escape the poisoned stream that passes via the male seed from generation to generation in all who are natural-born. In a way, we owe this discovery to the brilliant insights of August Weismann a century ago, whose failing vision forced him to forsake the microscope and the laboratory, and spend his working hours reflecting upon what he had already observed while his vision had been adequate.

The work of his successors has remarkably confirmed his initial hypothesis that in each generation it is the female seed that first reproduces itself and *then* forms the body which is to house it. The order here is crucial to a proper understanding. The succession of bodies are temporary vehicles which death lays aside – but not until the seed in the next generation has had time first of all to reproduce *itself*, and then to repeat the rest of the cycle. The body is the ovum's way of perpetuating itself.

Tracing this process backwards to Eve, we are forced to go one step further and say that Eve received her seed from Adam. There is thus a continuity of the seed originally in Adam, from one generation to the next, which still remains intact after all these thousands of years, "a bundle of immortality" which was once in Adam's loins. In each generation it is the now mortalized male seed that introduces mortality to the immortal seed of the woman – as

THIS DIAGRAM REALLY SHOWS TWO THINGS: FIRST HOW THE WOMAN'S SEED, ONCE IN ADAM, FORMED THE BASIS OF THE BODY PREPARED FOR THE LORD. SECONDLY IT SHOWS THAT IT IS THE SEED THAT FORMS THE BODY, NOT THE BODY THE SEED AS EVOLUTIONARY THEORY MANIFESTLY DEMANDS.

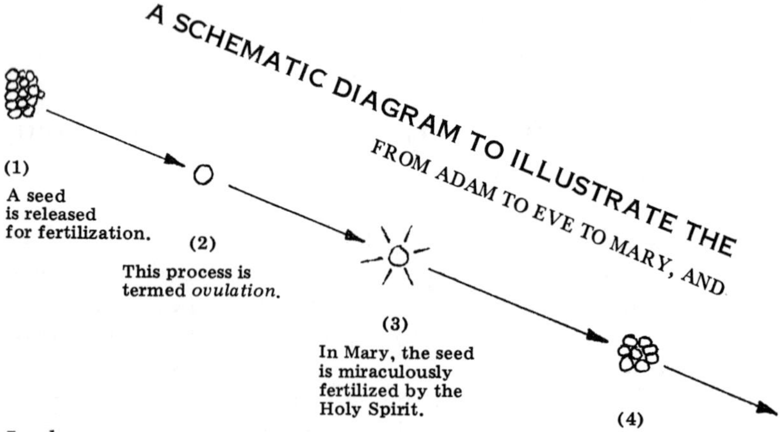

Luther and Calvin both perceived. But this fatal poisoning evidently does not take place until the woman's seed has multiplied itself and made adequate provision for the next generation by constructing also a body to house it. This process is shown above.

Thus while Eve became the mother of all living (Gen. 3:20), Adam had become the father of all dying (Romans 5:12). This is precisely stated in verse 12 where we are told simply, "by one *man* . . . . death passed upon all men".

And then one day, by divine intervention, the Holy Spirit introduced into the woman's seed in the virgin Mary that which initiated its development into a man-child, and by so doing, for the first time in history a woman was found to be carrying in her womb a "clean thing".

Luke 1:35 tells us that the angel said to Mary: "That *holy thing* that shall be born of thee shall be called the

Son of God". The rendering "holy thing" is perfectly justified by the original Greek and is by implication re-affirmed in Matthew 1:20 which, rendered literally would read, "for *that which* (neuter) is conceived in her is of the Holy Spirit". Once again the emphasized words are faithful to the original Greek.

When Mary's time was fulfilled, she brought forth a son, and the angels announced to the shepherds in the field, "Unto you *this day* in the city of David is born a Saviour who is Christ the Lord" (Luke 2:11).

That day the Lord of glory, the Son of God, became flesh and began to dwell among us as the Son of Man. And the Father in heaven confirmed the event, saying, "Thou art my Son, *this day* have I begotten Thee" (Heb. 1:5) – or as the New International Version has rendered it, "You are my Son, today have I become your Father". See likewise in the wording of Heb. 5:5

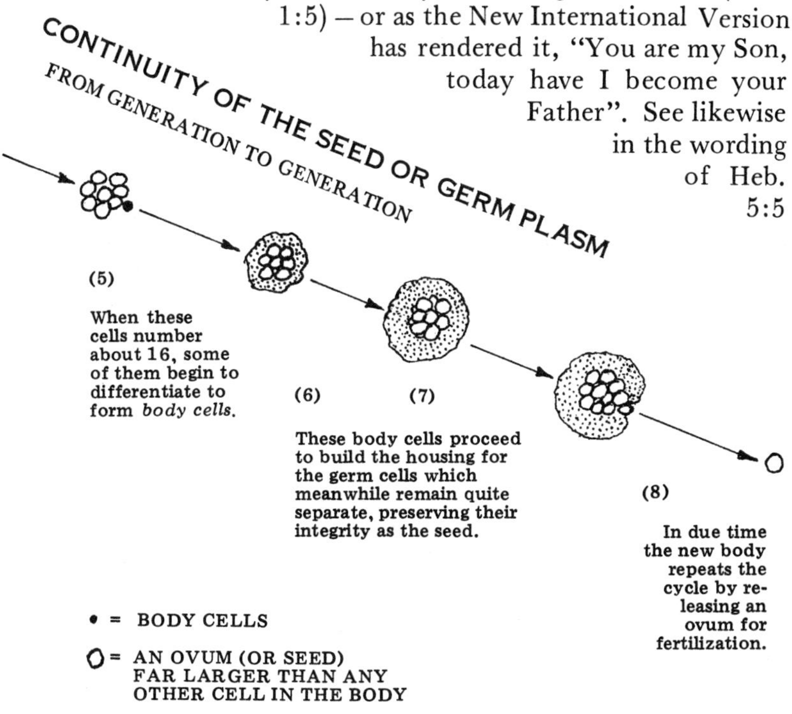

3. The recovery of a truly Adamic body.

"Wherefore when He cometh into the world, He saith . . . . 'A body hast Thou prepared for Me'." (Heb.10:5).

In the Greek of Hebrews 10:5, the word which is rendered "prepared" is a particularly significant one in the present context. Basically it means to *reconstitute*, to *restore*, even to *repair* rather than simply prepare. It is found in Matthew 4:21 in connection with the *mending* of nets. Moreover, the Greek Papyri show that it was currently used to mean to *prepare to perfection*. In Classical Greek it means to *furnish completely*. In the King James Version it has the meaning of *to perfect* – as in Matthew 21:16; Luke 6:40; 1 Thess.3:10; and elsewhere.

The idea in Hebrews 10:5 seems to be to underscore the fact that in some way this was a body that restored perfection in the Adamic line, contrasting his own body and its clean-ness with the un-cleanness of the bodies of all others hitherto born in this line. *His* body was flawless, and holy even in its foetal development and thus resolved the problem raised in Psalm 51:5 and Job 14:4.

Thus we may conclude without hesitation that the virgin birth did indeed produce a unique human body, truly Adamic in origin since the woman's seed was once Adam's seed, but free of all that fallen Adam has entailed to the rest of his descendants without exception. Here, then, was a perfect human body brought forth with the same potential for immortality that had characterized Adam's body as created.

The body of the Lord Jesus Christ was not therefore brought into being under the condition of our fallen bodies which come forth under sentence of death, but "after the potential for unending life" (Heb.7:16). Here was a *Second* Man, biologically fulfilling precisely the conditions which had characterized the body of the *first* man, which Augustine spoke of so perceptively as being *non imposse mori, sed posse non mori*: i.e., not unable to die (because He was vulnerable to the assaults of both the physical world and of men): *but able* **not** *to die* (because no mortogenic factor had ever been introduced into his body).

For Him there was no necessity of death. Unlike ourselves, therefore, his body was raised uncorrupted, still identifiably as his and *needing no change in it* save only that which transformed it to a new working principle to fit it for its heavenly role. Whereas for us there must always be a change through death – save only in the case of those who remain alive at the Lord's second coming again, who will somehow have a like change to fit their bodies for heaven (1 Cor.15:51).

4. How the New Testament has used a group of words to make clear the uniqueness of his body.

Let me illustrate what seems to me one very important way in which the New Testament has recognized a vital distinction between his body and ours, his temptations and ours, his death and ours, and so on. These distinctions have been blurred in most English translations. In the original Greek they are marked carefully by the use of two different classes of words, some of which are spelled with a prefix ending with an *i* (called an *iota* in Greek) and the others without the *i*. The first group of words is prefixed by *homoi-* and the second by *homo-*. Words prefixed by *homoi-* signify "likeness' with the sense of similarity, but the words prefixed by *homo-* signify "identity" or "exact sameness".

If I wanted to say "Margarine can look like butter" in Greek, I would have to use a word prefixed by *homoi-* for the English word "like", because margarine only looks like butter. It isn't actually butter at all.

In English we use words which have the prefix *homo-*, such as homology, homogeneous, homosexual, homonym, homozygote, etc., to mean identical structure, identical quality or consistency, identical sex, identical name, identical genes, and so on. *Homo-* conveys the idea, therefore, of precise identity, not merely likeness in appearance.

On the other hand, we do not use many words in English with the prefix *homoi-*. It is hard to say why this is, and it contrasts strongly with Greek usage both in the New Testament and in Classical literature. In Greek, words beginning with the prefix *homoi-* always signify mere similarity rather than precise identity.

Wherever words prefixed with either *homo-* or *homoi-* are used in the New Testament it is incumbent upon the translator to indicate to the reader whether the meaning is absolute identity or mere similarity, since great care is taken in Scripture in the distinctive use of these words. And this ought to be reflected in any translation. The distinction is always of quite crucial importance, but unfortunately many of even the best translations have failed in this respect because they have used the word "like" and "likeness" imprecisely. They have ignored the care taken by the divine Author of Scripture to mark a fundamental distinction.

Let me give a few specific illustrations from familiar passages in the New Testament where the difference is often masked in the translation but is in fact of great importance.

In Romans 8:3 we have the words, "God sending his son *in the likeness of* sinful flesh". In the original Greek the word rendered "likeness" is *homoi-omati*, and for this the word "likeness" is a correct translation. It will be noted that the prefix is *homoi-*, the *i* being part of the prefix. All such words signify *likeness only and not identity*. Thus the Lord's flesh was similar to ours but not identical. It was identical only with the flesh of *unfallen* Adam but by no means identical with the despoiled flesh of ourselves who are Adam's fallen descendants. Our bodies are corrupted (1 Cor.15:53); his body was not (1 Pet.1:18, 19).

In Hebrews 2:17 we have the words, "in all things it behooved Him to be *made like unto* his brethren". Had He been identical with his brethren, his body would have been mortal like ours, and He would have been under sentence of death as we are. The sacrifice of a body that is already under sentence of death can only be *premature* and never truly vicarious.

In Hebrews 4:15, "(He) was tempted in all points *like as we are*, yet without sin". The fact is that when Satan comes to tempt *us*, he comes to a citadel that has already surrendered. The root of sin is there to begin with, and Satan has only to appeal to it to find a ready accessory. When Satan came to tempt Jesus, he found nothing in Him to seize hold of, by which to work from within (John 14:30).

The Lord was always tempted from without: we are tempted from within. Indeed, we do not need Satan to tempt us, our own fallen nature being usually sufficient unto itself. The Lord was only tempted when Satan came to tempt Him: never otherwise. His temptations were every bit as real as ours but never arose from internal promptings.

In Romans 6:5, "We have been planted together *in the likeness of* his death". His death, and our deaths as individuals, are entirely different in that He died for many men's sins but without obligation for any of his own. When we are counted to have died in Him, we died in Him for our *own* sins. The element of vicariousness in our death is entirely absent.

And in Philippians 2:7 we read, "And was made *in the likeness of* men". The point in each of these important passages must by now be clear. Had He been made as we are made, conceived and born in sin, the consequences for mankind and indeed for the Universe would have been disastrous. For the human experiment could only have proved pointless without a Saviour – and conceived and born as we are, He could never have been a Saviour.

The second group of words has the prefix *homo-*, without the terminal *i*. There are some 46 instances of the use of such words, and always without exception the meaning is "identical with": not merely similar to, but *precisely the same*.

In Classical Greek literature the distinction between the

prefix *homo-* and *homoi-* is faithfully preserved in many verbs and nouns which, however, do not appear in the New Testament.

One of the historically most critical cases involving a compound word which can be prefixed by either *homo-* or *homoi-* appears in the formulation of the Nicene Creed (325 AD) in which the Lord Jesus was held by one party to be "of *one* [i.e., identical] substance with the Father" (*homo*-ousios), and by another party to be only "of *like* substance with the Father" (*homoi*-ousios). Some said that He was actually one with the Father: others said that He was merely like the Father. The Eastern and Western branches of the Church split over the difference between the prefix *homo-* and *homoi-*, or more precisely over the absence or presence of the *i*. It seems a foolish thing that Christendom should break in two at a critical point in its early development over the presence or absence of a single letter. But of course it was really over two entirely different concepts, mere similarity, or *absolute equality.**

This *i* is called in Greek an *iota*, and in Hebrew is termed a *jot*, the two words being cognate. It is significant, therefore, that the Lord should have said that no part of his Word should fail, not even a *jot* (Matt.5:18) until all had been fulfilled.

We conclude that when we are told the Lord Jesus Christ was made in the likeness of sinful flesh, or was made in the likeness of men, or was made like unto his brethren, we are to understand that what we have in this likeness is

---

* The following references are to words or phrases incorporating the prefix *HOMO-*. To promise faithfully: Matt.14:7. To confess plainly: Matt.10:32 (2x); Lu.12:8 (2x); John 1:20; 9:22; 12:42; Acts 23:8; 24:14; Rom.10:9, 10; Heb.11:13; 1 John 1:9; 4:2, 3, 15; 2 John 7. To profess forthrightly: Matthew 7:23. To be truly thankful: Heb.13:15. To be of the same craft, not merely a related one: Acts 18:3. Without doubt: 1 Tim.3:16. Together as one: John 4:36; 20:4; 21:20. Sincere profession: 1 Tim.6:12, 13; Tit.1:16; Heb.3:1; 4:14; 10:23. Manifestly declared: 2 Cor.9:13. In full agreement: Acts 1:14; 2:1, 46; 4:24; 5:12; 7:57; 8:6; 12:20; 15:25; 18:12; 19:29; Rom.15:6; and 1 Pet.3:8.

only similarity, not absolute identity. To have been identical with us would have placed Him under the same sentence of death that we are under.

In one important respect the Lord's body *was* identical with ours, simply because his was a physical body and therefore vulnerable. His body was just as subject to injury as ours. Like us He suffered fatigue, hunger, thirst, pain, and wounds – all such things as result naturally from the demands of any physical body, whether human or animal. These all come under the first part of Augustine's aphorism: He *could* experience death. In this sense, He was exactly as we are. We are told in 2 Corinthians 13:4 that He was crucified through weakness. This was not the weakness of sin but the vulnerability of a real body, as it was of *unfallen* Adam's body. It was no sign of sinfulness that He could be wounded for our transgressions. Ambrose (c.339 to 397 AD) who had such a powerful influence on Augustine, wrote:

> Thus it is written, "God made this Jesus whom ye crucified both Lord and Christ" (Acts 2:36). It was not the Godhead but the flesh that was crucified. *This indeed was possible because the flesh allowed of being crucified*".

This "infirmity of his flesh"* was not an infirmity due to the effects of sinful flesh, but the vulnerability of all things set within the framework of the physical world.

Scripture has provided us with enough information as to how this uniqueness of his body came about. From the very beginning, nature was designed to make all this possible without violation of its own order; to perpetuate unfallen Adam's constitution and avoid the entailed damage from the Fall; to produce as an end result a unique embryo which unlike all other human embryos was "clean"; and

---

* Cf. *Belgic Confession*, Article XIX: "very man that He might die for us according to the infirmity of the flesh".

thus to re-introduce into the world a Second Adam whose body was not subject to death and *yet could experience it*: none of these achievements violated nature as God designed and created it. Into this body, perfectly prepared for Him, the Son of God came to be our Saviour as the Son of Man, becoming what He had not been hitherto yet without ceasing to be what He was before. Only thus could the Plan of Salvation by substitutionary sacrifice be made effective.

In short, the divine nature was in no wise demeaned by the assumption of a perfect human body. Such, then, was the form and dignity and capacity of the body with which Adam was created. In the Lord Jesus Christ true manhood, body and spirit, was once again displayed in all its immortal glory before a fallen world.

The divine Logos who was with the Father throughout all eternity and through whom the Universe was created and by whom it is kept as a Cosmos rather than a Chaos, became Man and dwelt as a Man among men, and we beheld his glory, the glory as of the only begotten of the Father, full of grace and truth (John 1:14).

Everything hinges upon the perfection of that body; everything hinges upon its being truly human; everything hinges upon its being vulnerable; and everything hinges upon its being contingently immortal.

Such a glorious house was his!

It is conceivable that as it now is, man's body might be accounted for by evolution. But in accounting for the body of Adam as witnessed in the body of Jesus Christ, fulfilling these four prerequisite conditions, evolution utterly fails to help us at all.

Chapter 12

THE INVISIBLE BECOMES VISIBLE

The mystery . . . of God . . . manifest in the flesh	1 Tim.3:16
The glory of God in the face of Jesus Christ	2 Cor.4:6
In whom dwelleth all the fulness of the Godhead bodily	Col.2:9

It was not long after William the Conqueror crossed the Channel from Normandy and landed on the English south coast in 1066 that our friend, Anselm of Canterbury, wrote his famous treatise on the Incarnation to which we have already referred. You may recall that he titled it, *Why God Became Man*.

His answer was essentially what we would say today – "to become the Redeemer of Man". Many of his arguments are a delight to read and satisfying to both heart and mind. But there were some questions he did not address. Moreover, the answer which he did give, and which most of us would give, is not by any means the *only* reason why the Eternal Son of God became Man and dwelt among us.

There are in fact a number of reasons beyond the redemption of man, of which the following three are worthy of special attention in the present context, and surprisingly all three of them required *embodiment*.

(1) He came to reveal God to man.
(2) He came to reveal man to God.
(3) He came to reveal man to himself.

In this chapter I want to address only the first two, and then deal with (3) in the next chapter.

(1) He Came to Reveal God to Man

The renowned British historian, Arnold Toynbee, was one day discussing man's knowledge of God with a close friend named Dr. Edwyn Bevan. During the conversation Bevan said to him, "Man's vision of God is like a dog's vision of his master. The dog by instinct, habit, and association, comes to know his master in a limited manner. But to know him fully the dog would have to forsake his canine nature for a human nature".

Now, I have no knowledge of how the conversation proceeded from that point, but it naturally started me thinking about the Incarnation in that light. Of course, for the dog to become a man is a reversal of the situation in which God became man. For we have, in the first instance, the lesser becoming the greater whereas in the latter case we have the greater becoming the lesser, the infinite becoming the finite. Bevan does not here refer to the possibility of the man becoming a dog in order to understand his faithful pet, though it seems the natural alternative to raise for discussion.

But does Bevan's solution really solve the problem of communication in any case? Suppose the dog becomes a man, would he not then of necessity *cease being a dog*? The dog who has become a man is no longer a dog! If he ceases to be a dog, what good is he to the man as interpreter of a dog's thoughts? In the very act of becoming a man he loses all contact with his former canine nature. On the other hand, to go halfway and become half a dog and half a man is to be neither dog or man, and in this kind of neutrality such a creature cannot wholly reveal the dog to the man or the man to the dog.

The difficulty becomes a very practical one when anthropologists or missionaries employ a native interpreter. The interpreter must already have escaped his own culture in part and immersed himself in the alien culture of the anthropologist or missionary in order to be a good interpreter. How can he become a good interpreter without adulterating his understanding of his *own* culture in the process of learning to interpret it in terms of the alien culture? He is no longer "pure" native. Who knows thereafter how much of what he tells his enquirers is genuinely native and how much is unconsciously adopted from the alien culture?

If there is no real solution at all by this route to the problem of communication, wherein does the answer lie? What form must such a 'bridge' between God and man take which does not surrender one or other nature? Can you build a bridge with the two ends in the middle? How can God possibly become man while yet remaining God? How did it come about that the Son of God could really become the Son of Man without ceasing to be what He was before? It would seem to be a sheer impossibility for the man to become a dog and remain a man. How, then, did God become a Man while remaining God?

The answer lies perhaps in this: that the *nature* of God and the *nature* of man as originally created, shared a certain fundamental compatibility, which man and dog do not share despite the real companionship that may exist between them.

Because man was originally made in the image and likeness of God (Gen.1:27), his nature was compatible with the divine nature, thus making it possible for the divine nature in the Person of the Son of God to be fashioned in the likeness of man (Phil.2:7) while at the same time retaining the precise image of the Father (Heb.1:3). For this reason the Son of Man was able to mediate to us the capacity to partake of, even in our present state, something of the divine nature (2 Pet.1:4).*

---

* As, in the Incarnation, God partook of the nature of man without ceasing to be God, so, when redeemed, man may now partake of the divine nature — *without ceasing to be man.*

This duality of his nature constantly broke through to his companions and to his critics, causing the former to worship and the latter to condemn. On one occasion He had healed a man "with an infirmity of thirty and eight years" on a Sabbath day (John 5:5-9). When accused by the Pharisees of breaking the law by "working" on a holy day, Jesus replied, "My Father worketh hitherto, and so do I" (verse 17).

This was seized upon at once as blasphemy by the Pharisees who took up stones to stone Him. When He asked them why it was blasphemy, they replied, "Because thou, being a man, makes thyself God". They argued thus because Jesus had said not merely that God was his Father but that God was *his very own* Father. For this is how the Greek actually reads.

In point of fact, the Pharisees were doubly in error! For it ought not to have been said of Him that He, being a man, was making Himself God. Quite the reverse: it was rather that He, being God, had made Himself man!

Now, for over three years the disciples had walked with the Lord, rubbed shoulders with Him, seen Him daily performing wonders worthy of God Himself, while at the same time responding to the ordinary circumstances of life precisely as any other man would have done. He was often tired, sometimes hungry and thirsty, and in a multitude of ways humanly vulnerable, so that He had occasionally to escape from the crowd to protect Himself. Everything conspired to place the stamp of common humanity upon Him, and yet out of that common humanity there kept breaking through something that shook the disciples and made them wonder what kind of Person He really was.

One day, feeling hungry like any other person might, and seeing a fig tree a little way off which was displaying the characteristics of a tree bearing fruit despite the fact that it was not the season, He went eagerly towards it with every expectation of finding something to eat. Both his natural hunger and his reasonable expectations were normal to any man. However, He was disappointed: there was no fruit on the tree. Humanly speaking, He had been deceived.

For reasons which are not altogether clear, though perhaps because He desired to make the point for the disciples that a false witness was to be condemned, He simply decreed that the fig tree should no more bear fruit thenceforth. His power to do this was quickly confirmed, since by the very next morning the tree had already withered — to the amazement of Peter (Mark 11:21).

Here was a striking case of what looks like a contradictory co-existence in one person of a human nature subject to hunger and surprise, with a divine power over inanimate forms of life that was absolute.

This kind of juxtaposition was observed so frequently by the disciples that it dawned upon them they were indeed in the presence of some One quite different from and yet strangely the same as themselves. He seemed to be sometimes limited in his knowledge and at other times omniscient. On one occasion Philip said to Him, "Lord, show us the Father and we'll be satisfied". Perhaps they were *always* asking Him questions. Jesus replied to him, "Have I been so long a time with you and yet you have not recognized who I am, Philip? He that has seen Me has seen the Father; and how can you say then, 'Show us the Father'?" *

Thus by repeated demonstrations of the reality of his dual nature, human and divine, He had been making visible

---

* Theodoret (393-458) in one of his *Dialogues* says this to one of his (fictitous) antagonists: "How then was it possible for the invisible nature to be seen without a body? Or do you not remember those words of the Apostle in which he distinctly teaches the invisibility of the divine NATURE? He says 'Whom no man hath seen nor CAN see' (1 Tim.6:16)?" Theodoret is speaking of the divine NATURE, and strictly speaking, the *nature* of nothing is visible until it is *objectified* in some way.

Leo I the Great (400-461) wrote: "The Son of God therefore came down from his throne in heaven without withdrawing from his Father's glory, and entered this lower world, born after a new order by a new mode of birth. After a new order, in as much as He is invisible in his own *nature*, and He became visible in ours [i.e., in HUMAN terms, and showed God to be person-al]; He is incomprehensible and He willed to be comprehended; continuing to be before time He began to exist in time".

in his Person the invisible Father in heaven whom man could not otherwise have seen. These sudden juxtapositionings of the human and the divine in the Lord Jesus Christ were constantly being displayed, but with such simplicity that the disciples were only surprised at the striking elements in the latter because they had no difficulty whatever in accepting the reality of the former.

Take the familiar case of the terrible storm that arose on the Sea of Galilee, when Jesus fell asleep on a pillow in the stern of a small fishing boat. The event is noted by all three Synoptic Gospels, a circumstance which suggests that it left a profound impression upon them all. Mark 4:35-41 provides one of the simplest accounts.

There is a wonderful correspondence with life in the Word of God. No matter how weary a man might be, it is doubtful if he could sleep very soundly in the *bow* of such a small vessel on a choppy sea, and he would be subject to almost constant wetting by the spray. So we are told that He was asleep in the *stern*. In modern vessels with a sternpost rudder such a place to sleep would be most inconvenient for the helmsman, but in those days there were no such rudders, only a board over one side to steer with. It was called a steer-board and gave rise to our word *starboard*.

On this occasion, while the Lord had fallen into a deep sleep there had arisen "a great storm". Most small hill-bound inland seas and lakes are subject to such sudden storms, and all too frequently they are remarkably violent. The Great Lakes of Canada and the United States are notorious for their wicked behaviour. Ocean sailors who know these lakes are very respectful of them.

Over 6000 ships are recorded to have sunk in these lakes, and these ships were by no means small. In one single terrible night (November 9, 1913) over 30 ships were wrecked, 10 of which sank without a trace with all hands lost. The roster of wrecked vessels included ships of 269 feet in length, 270, 440, 452 and 524 feet, and most of them steel-hulled. This was on Lake Huron which can be vicious because it is shallow, though Lake Superior can be even more disastrous for the crew because the water is so cold. On that one memorable night waves of 60 and 70 feet in height crashed over and swept across the decks and wiped them clean of bridge, deck housing, funnel, and crews in their stern quarters – everything. Winds can be cyclonic and in this case persisted for 16 hours without a break, making man and his machinery utterly helpless.

Many travellers in the Middle East have observed the same sudden violence on the Sea of Galilee despite its small size (6 miles x 12 miles). Evidently the storm to which Mark 4 refers was such a storm, so sudden as to take even the experienced fishermen by surprise. They were very soon in real danger of sinking – and yet the Lord slept on. How *human* was such a total weariness as this!

The disciples awoke Him in desperation and appealed to Him: "Master, carest thou not that we perish?" What did they really expect Him to do?

Without a word of rebuke to *them*, He arose and, instead, rebuked the wind and the sea, saying, very simply, "Peace, be still"! And the wind ceased, we are told, and there was a great calm. The sudden silence must have been almost shattering. Then, and then only, did He rebuke the disciples for their lack of faith, for was He not with them in the boat? They in their turn must have been exceedingly relieved but also truly fearful, for they said to one another, "What kind of man is this, that even the wind and the sea obey Him?" (Mark 4:41).

What kind of a man indeed, if He was not also God? But then *we* know; He was acting as the Lord of the Old Testament, since this storm is described in Psalm 107:23-30. It is here recorded in extraordinary realism.

> They that go down to the sea in ships, that do business in great waters; these see the works of the Lord, and his wonders in the deep. For He commandeth and raiseth the stormy wind, which lifteth up the waves thereof. They mount up to the heav-

> ens, they go down again to the depths: their soul is melted because of trouble. They reel to and fro, and stagger like a drunken man and are at their wit's end.
>
> Then they cry unto the Lord in their trouble, and He bringeth them out of their distresses. He maketh the storm a calm, so that the waves thereof are still. Then are they glad because they are quiet.
>
> So He bringeth them unto their desired haven (i.e., home again).

Surely this was a prophetic forecast of this very storm, for it was the same Lord who with absolute authority stilled the waves by a single word of command.

Here we have, then, at one moment the Lord Jesus asleep as only a terribly tired man could sleep – *for it is impossible to imagine that He was pretending;* and the next moment the same Lord is commanding the winds and the waves to cease – and they obey so instantly and so literally that the disciples are truly amazed and ask themselves What kind of *man* is this.

Let us take one more instance of the same juxtaposition of truly human and truly divine behaviour in a single individual. The details are given in John 11:1-44.

Lazarus, after a terminal illness that seems to have been very brief, died and was buried in a garden tomb in Bethany. This Lazarus and his two sisters, Martha and Mary, were all three of them particularly loved by the Lord, as John 11:5 makes clear. If the Lord did have a 'home' on earth, it was with these dear people. Nevertheless, He did not at once respond to their call to come to their help, and He arrived too late to save Lazarus from death – although it is clear that He could have done so.

When He did arrive, Lazarus was already four days dead and buried, and the process of putrefaction had begun. Almost at once it would seem, Jesus enquired where they had laid him, thus displaying the natural ignorance of any other man in such a circumstance. Then together they walked towards the burial place, and as they walked Jesus Himself was overcome by the grief of his two beloved friends and could not restrain his own tears.

We thus have, once again, two clear evidences of a truly human nature marked by limitations of knowledge and incomplete emotional control. But the moment they reached the tomb the divine nature asserted itself. He commanded those who stood by to roll away the stone.

At this command, Martha at once protested! What terribly disfigured spectre would the light of day reveal in that place of death and decay? Current means of embalming did little to preserve the body. The thought of exposure must have horrified her: "Oh no! Lord, by now he's . . . . he . . . . the odour, Lord! He has been dead four days already".

We do not know how the Lord reassured Martha and Mary but without hesitation He signalled to those who stood by the stone to proceed, and they had soon complied with his command. Then Jesus cried out in a loud voice, "Lazarus, come forth!"

And he that was dead came forth from the tomb, wrapped hand and foot and head – separately, as was customary. And there he stood in the opening of the tomb, his face and his body still concealed. Imagine the fearful look that Martha and Mary must have cast towards this apparition. And imagine their even greater terror when Jesus said, "Unwrap him and set him free".

What would they see beneath those wrappings? Some kind of ghastly travesty of a human face fit only for a horror movie: or the face of a dear soul as they had known him before his illness? Whatever they might have hoped before, was now almost certainly overbalanced by what it was natural to expect after so long in the grave.

We are not told what did happen when they recovered their brother, healed, made whole, and alive again! Nor are we told what the Lord Himself did afterwards. Did He walk back with the three of them to the house? And what could they talk about on the way? The curtain is discreetly drawn, for Scripture never concerns itself with the satisfaction of mere curiosity. For the present, that was the end of the matter.

He who had only shortly before told Martha that He was the resurrection and the life, a claim which only God could make, and later had wept, which only man does, had now proven that his higher claim was true. He was indeed the Son of God and the Son of Man: not two *Sons* but having two *Sonships* in his one Person.

Thus in the Lord Jesus Christ we see a manifestation of the nature of God, God "objectified", the invisible made visible. From this revelation we see above all that God is personal: not some mighty force, but a Person. We observe the reality of this person-hood in the very range of reactions of the Lord Jesus Christ. The Lord was hungry in the wilderness (Matt. 4:2); thirsty on the cross (John 19:28); bone-tired at noon beside Jacob's well (John 4:6); asleep on a pillow on a stormy sea (Mark 4:38); overcome with grief (John 11:35); tortured in body with frightful wounds at the hands of Pilate's soldiers (John 19:1-3); and physically exhausted (Matt.27:32) by his suffering on the way to "the place of the skull". All these were evidences of a truly human Person in a perfect body unmarred in any way by sin. And yet this same human Person could command plants to wither, storms to cease, loaves to multiply, water to become wine, and the dead already putrefying to come forth whole.

If God the Father is revealed here, then how wonderfully personal our Father in heaven really is. You might say He is almost human! Indeed, more human than we are. It is not surprising perhaps that Karl Barth could write a book on *The Humanity of God*. How otherwise could He have created man in his own image?

Thus was fulfilled one purpose of the Incarnation: a

revelation of the personal nature of God and a demonstration of his great concern for man. God also is touched with the feeling of our infirmity, even while He is truly angry at our sin – just as the Lord Jesus was on many occasions (Mark 3:5), even in those whom He loved (Mark 10:14).

2. The Revealing of Man to God.

Does it really make sense to speak of something that had to be revealed to God? Could there possibly be such a need? Above all, is there anything hidden from God that He should need a Man to reveal it to Him? Surprisingly, there is indeed. And it all hinges on embodiment!

What had to be revealed to God was *the nature of human temptation*. It is that form of temptation to which the vulnerabilities of human embodiment have exposed man. Such temptations cannot possibly be experienced by a purely spiritual being such as God is. Yet these demands of the body are enormous.

God cannot ever have known by experience temptation due to hunger, or thirst, or physical pain, or weariness of the flesh, even less the fear of physical death which plagues man for most of his life.

How, then, was God to judge man who is subject to so many temptations that acquire much of their force from the demands of the flesh? As we have noted, many of these demands do not stem from *sinful* flesh per se, but rather from the mere fact of bodily existence even in a house as perfect as was the body of the Lord. He, too, suffered from hunger and thirst and pain and fatigue. Even in his perfect body these could have been – and in some cases we know they were – avenues of attack by Satan.

It was Satan who tempted the Lord to appease his hunger in the wilderness: and unless He *was* hungry, there could have been no temptation. But this was true also in the matter of thirst, as we have already seen (page 30).

But as for the Father in heaven, how could He ever

know what drives man to do some of the things he does – to steal because starving, to fight because of thirst, to lie because of unbearable torture, to fail in prayer because of sheer fatigue, to drink when the body makes demands beyond the spirit's bearing, and to over-indulge because the appetite is stronger than the will.

Our lives are neither purely spiritual nor purely physical so that there is often a conflict between the two and we are torn between different kinds of impulse in ways which can never be experienced even by an angel. And what about the temptations which arise from our bondage to time and to place simply because we cannot wait?

Because we belong in a physical world as well as a spiritual one, a purely spiritual being knows only half of what we are subject to by way of temptation. Is it not therefore only proper that the Father should send his Son to share our experience in the fullest way possible short of involvement in our sinful nature and its potential? Is it not reasonable that He would seek to "know" through his Son how our very physical existence contributes to our fallenness? And then having received his Son back again unspoiled by sin as man has been but only made even more mature in his manhood by the things He experienced (Heb.5:8), to delegate to *Him* the office of Judge of Man, *in his own stead*?

In the old days when society was highly stratified by a class structure, one of the provisions of the law was that a man ought if possible to be tried and judged only by his peers. It was held to be unfair for a man in a certain situation in life to be judged by someone who could not know anything about that kind of life by experience, and would therefore be largely ignorant of the nature of the condemned man's temptations. Of course the justice of the class system can indeed be called into question to begin with. But the idea did have a measure of fairness about it, given the realities of social structure at the time.

In scientific circles today, we feel that the value of a

man's work can only be fairly judged by someone in a position to assess it knowledgeably, and not by one who has no such background experience or competence. We call this a system of "peer review". It is, of course, like all else that man does, far from perfect in its operation; but anything else could be considered entirely unjust.

Since God cannot be tempted *at all* (James 1:13), and thus cannot know by experience even the meaning of being tempted except by observing its effect upon the individual himself and on others, how could his judgment of human behaviour be entirely just? Perhaps it would be better to say, How could any such judgment be *seen* to be just if the Judge Himself knows nothing personally of what temptation means under such conditions? A just trial demands that it not only *is* fair, but also seems fair to those who witness the proceeding. Will the Judge of all the earth do less than we attempt to do?

At any rate, whatever the divine 'rationale', this at least is clear: all judgment has been assigned to the One who became Man without ceasing to be God. As the Lord said plainly: "The Father judgeth no man, but hath committed all judgment unto the Son . . . . and hath given Him authority to execute judgment also" (John 5:22,27). That is to say, not merely to pass judgment but also to carry it out. And why this delegated authority? Again the answer is straightforward: "*Because He is the Son of Man*" (verse 27). That is the specific reason, because He *is* the Son of Man.

Paul fully confirms this fact in Romans 8:34, "Who is he that judgeth? It is Christ who died, yea rather, who is risen again, who is even at the right hand of God; who also makes intercession for us". Thus the very One who is the Saviour is also to be the Judge and to plead our case!

And again in Acts 17:31 we find the same insistence that it is not merely One who lived and died as a man but who was also raised as a man that is to be the world's Judge.

"Because (God) hath appointed a day in which He intends to judge the world in righteousness, by that MAN whom He hath ordained: whereof He hath given assurance unto all men, in that He hath raised Him from the dead".

In 2 Corinthians 5:10 Paul writes that we, the children of God, are also to come before the judgment seat of Christ that we may be declared worthy of praise for the good that we may have done, and thankfully see our failures removed from the record through his mercy and for his own name's sake. And Paul again lays emphasis on the fact that it is a judgment of "things done in the body", done by us not as spiritual beings but as embodied beings.

This emphasis was often underscored by Tertullian whom we have already quoted in another connection. In his treatise, "On the Resurrection of the Flesh" (chap. xiv), he observed:

> Thus it follows that the fullness and perfection of the judgment consists simply in representing the interests of the *entire* human being. Now since the entire man consists of the union of two natures [i.e., both the physical and the spiritual], he must therefore appear in both, since it is right that he should be judged entirely; nor, of course, did he pass through life except in his entire state.
>
> As therefore he lived, so also must he be judged. because he has to be judged concerning the way in which he has lived. For life is the cause of judgment and it must undergo investigation in as many natures as it possessed when it discharged its vital functions.

Tertullian therefore concludes that the flesh ought not to have any share in the sentence, either for praise or for blame, if it had no share in the cause of it. Since it clearly is to come into judgment, then obviously it *did* have a share in the cause.

Now since *the Father* has never experienced embodiment, accordingly it was essential for the Son, if He was to become our Judge, to experience human embodiment because that embodiment plays such a part in the causes of our condemnation. He was therefore incarnate for this reason also: that He might reveal man to God, manifesting and communicating to his Father the nature of embodiment as it bears upon the nature of human guilt.

Just how the Father shared in the experience of the manhood of the Son has never been adequately spelled out, though many attempts have been made. Was it by a kind of empathy, such as moved the Lord to tears beside the tomb of Lazarus in spite of the fact that He knew He was about to undo the cause of it all?

All we can say is that the Father has committed all judgment to the Son because He was tempted as an embodied human being and therefore understands the human situation in a way that the Father never could. It would seem therefore that the Father has, as it were, withdrawn from that office in fairness to man and committed judgment to the only One who could exercise it justly.

It would be hard indeed for us to conceive of a way more just and fair. Could there *be* a more excellent way? In Him was the pure essence of manhood and the pure essence of deity. And He placed Himself voluntarily in the position of experiencing the worst that the world could do. He thus becomes the perfect Judge between fallen man and a righteous God.

The true bridge between Bevan and his dog could not take the form of a human spirit in a dog's body nor a canine spirit in a human body. Somehow no such bridge is possible, because the dog was not created in the image of the man. It would be neither dog nor man but a monster. But man *was* created in the image of God and thus God could and did become man by human embodiment while yet retaining his divine nature. It is clear that a human body did not make this amalgam in any way inappropriate. Thus the Lord Jesus as Son of God and Son of Man can stand in fairness as the Judge of men.

Among those to be judged in fairness because of this shared image is, of course, Adam himself, the first human being. Unless the Lord was made in the image of *Adam* and not merely in the image of we who are Adam's descendants, He cannot fairly judge Adam's temptations. Adam's body therefore cannot possibly have been the kind of primitive, barely human, body that evolutionary theory demands for the first man. The body of the First Adam as created must have been in every way homologous in form and function to the body of the Last Adam.

Chapter 13

TWO MEN: TWO ADAMS

The *first* Adam, as truly the *first* Man, ONLY SINNED ONCE.
The *last* Adam, as truly the *second* Man, NEVER ONCE SINNED.

Should you ask how the first Adam, as truly the first Man, only sinned once, then consider this fact. When Adam was created, he was created in the image of God. When he sinned, he surrendered that image and the specific nature that it signified. He literally sinned into being a new kind of creature, a species quite different from that which God had planned when He first said, "Let us make man" and then defined his creation by the words "in our image" (Gen.1:26). The first man Adam, as truly representative Man, committed only one sin because with that one sin he ceased to be representative Man. All his other sins are of no significance to us because they were not sins of Adam as Man judged by God's definition of the word *Man*. Adam's first sin as truly man was his only sin as truly man. Thereafter Adam sinned as a creature who was not truly man any more.

That this creation of God did surrender that image is borne out by two facts. First, because the sons and daughters of Adam and therefore all the descendants of the first man were no longer "made" in the image of God but in *Adam's* image. It will be noted that Genesis 5:1 states very specifically that God created man in *His* own image, but in Genesis 5:3 it is stated equally specifically that Adam begat sons and daughters in *his* image. And secondly, the New Testament indicates that God's image now has to be re-constituted (2 Cor.3:18).

It will also be noted that in Genesis 9:6 we are told that Adam WAS made, not that he IS made, in the image of God. The Hebrew of the original here is very specific and quite unambiguous. The past tense is used, not the present. Man lost that image when Adam sinned and in losing it ceased to be Man by God's definition. Yet, although man has indeed lost the image, he has not lost the capacity for its re-creation; for which reason the killing of a man is such a criminal offence, for fallen man has remained redeemable. Fallen angels did not remain redeemable: fallen man did.

But with the Incarnation of the Lord Jesus Christ a Man was born once again into the world bearing the express image of God (Heb.1:3); and He, unlike the first Adam, never lost that image because He never once sinned. Thus, Man as originally made in the image of God, still constituted the perfect vehicle for the revelation of God by incarnation, since by that Incarnation could be demonstrated God's love for his creature even though that creature had lost His image.

We thus find a sequence of events as follows:

Man was created in God's image:	Gen.1:26
Man sinned and lost that image.	
Man is now procreated in man's image:	Gen.5:3
God was made in man's original image:	Heb.1:3
Man may be re-constituted in God's image:	2 Cor.3:18

Now I want to address in this chapter a fact of history that evolution *cannot* account for. First: fallen man with all his destructive and suicidal propensities, and secondly: unfallen Man in all his consummate magnificence. Both Adams are truly representatives of *human* nature. How have these opposites come about? Do they really spring from a single root? What kind of a root could give rise to two such extremes, man as seen in ourselves and man as revealed in the person of Jesus Christ? Does not this signify something quite unique about the origin of that root?

1. FALLEN MAN.

On one occasion C. S. Lewis quoted a famous couplet from Bobby Burns' poem *Man Was Made to Mourn*.

> Man's inhumanity to man
> Makes countless thousands to mourn.

His comment on these two lines was very perceptive. He said the truth is that "it is not man's *inhumanity* to man that is the problem". No! It is man's "humanity" that is the problem. His awful behaviour is now "human behaviour". *This* is his nature. This is what man now is.

The popular view that man is essentially good dies hard. The concept of a steady improvement of human nature by education received a severe shock when two world wars showed that one of the most educated nations in the world could behave on an unprecedented scale in barbaric ways that were far worse than man had ever witnessed in terms of the numbers hurt by them and the depth of degradation to which they were subjected – and indeed in some countries are still being subjected.

It was Rousseau who had held up the noble savage and argued that here was a picture of unspoiled human nature which only civilization had destroyed. He advocated a return to such "native nobility"; and many have tried it. Not one of these returns to nature has resolved the problem of man's innate selfishness and the plague of a stricken conscience that remains to trouble the community and the individual alike.

Many studies of the Nazi concentration camps have been made since World War II. The incredible cruelties which were commonly inflicted or authorized with sadistic pleasure by people who then went home to enjoy fine art, classical music, and elevating literature, only go to show how terribly human nature has been warped by the Fall. Concentration camps and torture chambers are a *human* invention. The rest of nature displays nothing that could even be remotely viewed as the foundation of this.

The horror of those concentration camps was so awful that normal civilized people who witnessed them simply could not believe their own eyes. These were not visitors who thus reacted, but themselves victims of the horror. One doctor, seeing the lurid flames of a large fire some little distance away, wondered what was being burned – rubbish, he supposed. A truck backed up and men with ordinary pitch forks were tossing small bundles of garbage, one forkful at a time, into the flames. It was night and their silhouettes stood out like demons feeding the flames of hell.

Suddenly he realized what those small bundles of garbage actually were. They were babies, and he was certain that some of them were still alive – they were struggling on the tines of the fork. What happened to him, as he watched? He merely turned off; it was all a dream. It was simply relegated to fantasy. He knew it was true: he refused to believe what he knew . . . . Human beings are not capable of such actions. He would, he was sure, later find it was a dream.

Another scene. Women who became pregnant were treated mercilessly. They were kicked in the stomach, dragged by the hair or, worse, by one leg to the furnaces and after more physical abuse cast alive into the furnace. People who witnessed this, too, simply did not believe. Yet later they knew it was really done.

It seems that man alone of all creatures seeks to hurt his victim, deliberately, eagerly, furiously, viciously, *with incredible cruelty* – and with maniacal delight. Governments choose sadists to conduct their torture sessions and supply them with the "latest" technological devices, most of them intended for extreme physical abuse short of death. And the civilized nations market many of these devices. They are violent or slow and excruciating. They are applied to those parts of the human body which we consider more private and which are most sensitive. They are the most degrading devices in terms of the victim's reactions. But even so, perhaps human excrement plays the most terrible part of all . . . . Forced into the mouth, the nose, the ears . . . . forced as drink and food . . . . It is *incredible*.

All nations have been guilty, the civilized as well as the uncivilized. If we do not believe in demons it is only because we are so ignorant of what man can do to man when inspired by hatred.

William Temple was absolutely right when he said that the worst things that happen do not happen because of a few people who are monstrously wicked but because we all are what we all are. It is almost accidental that only a relative few in any society do these things. In the same circumstances the mildest of men can become worse than animals by far, for animals do not tear each other's flesh for the mere pleasure of hearing their screams.

We have found in the Nazi era an easy target and a ready source of illustration but they are no worse than we are *in potential*. Animals do not hate; only man does. And human hatred is inspired by the devil.

George Steiner was right when he said of those places of horror: "Art, intellectual pursuit, the development of the natural sciences, and many branches of scholarship, flourished in close spatial proximity to massacre and the death camps". It is a fact that these pursuits were being followed with devotion and enjoyed in such close proximity to these horror camps of pure bestiality: and they were fully *aware* of this proximity.

Aesthetic feeling, moral indignation, inventiveness, scientific competence, intelligent preparation for action, even concern for others – all these can be found sitting side by side, as it were, with such places of horror. So totally inconsistent is man's moral sense that an individual under oath will tell the most blatant lies in order to prove that he is innocent, a man of honour and integrity!

Such inconsistency is borne out by the fact that many normal and enjoyable neighbours living in the environs of these camps pretended not to know what went on. Yet many of these same people changed the furniture of their rooms to place the daily living quarters on upper floors so that they had a better view . . . . When asked why, their only answer was silence. We do not *know* ourselves; none of us really do.

It may be thought that only the Germans, "they" from our point of view, ever acted so atrociously. This is quite untrue. Dostoyevsky records an incident from Russia that is just as monstrous: the scale is smaller, but the phenomenon is the same.

One day, a serf boy, a little child of eight, threw a stone in play and hurt the paw of the local general's favourite hound.

"Why is my favourite dog lame?" he asked. He is told that the boy threw a stone that hurt its paw.

"Take him", he ordered. The child was seized from his mother and kept shut up all night. Early next morning the general came out on horseback, with his hounds, his dependents, dog-boys, and huntsmen, all mounted and in full hunting dress. The servants, too, are summoned for their edification, and before them all stood the mother of the child.

The child is brought out. It is a gloomy, cold, foggy autumn day – but capital for hunting. The general orders the child to be undressed and the child is stripped naked. He shivers in the cold, numb with terror, not daring to cry.

"Make him run", commands the general.

"Run! Run!", shout the dog-boys. And the child runs.

"At him!" yells the general, and he sets the whole pack of hounds on the child. The hounds catch him and tear him to pieces before his mother's eyes . . . .

But even this is "they" – the Russians, not us. Yet is there really any difference between the pleasure they derived from such utter brutality and the pleasure that the "professional" cock-fighters get out of their cruel sport, sanctioned in America in a number of States and fully protected by law? Is cruelty to animals, for pleasure, really any less an exhibition of man's innate fallen-ness? No animals do this to each other for mere amusement.

Such wickedness is everywhere in our own society. It is not overt and therefore is not so offensive to us, but it is there. The poor who abuse the welfare system, the lazy who abuse the unemployment insurance; the people who give "donations" that don't exist and receive an "official" receipt which is submitted as an income tax deduction and then share their tax savings with the non-profit organization that issued the receipt – all by prior arrangement.

The fact is that while some sin is so awful because it is public, most sin is private and therefore concealed or by many "overlooked". The fallenness of man is deep and wide, it is universal, for "all have sinned". It is only by accident that we personally may have escaped doing these more horrid things, because we were never placed in the position of being driven by hate or anger to do them – or if we were, we were not *able* to do them at the time of our anger or hatred. Only kings can traditionally do what they like, being a law unto themselves. David, Israel's *best* king, and Ahab, Israel's *worst* king, both turned coveting into murder. One coveted another man's wife and murdered her husband; the other, another man's vineyard and murdered its owner (2 Sam.11:1-27 and 1 Kings 21:1-16). This is what we are capable of, the best and the worst of us alike – *given the power.* David utterly repented, Ahab was utterly indifferent. Nevertheless, both behaved murderously because both were fallen men.

But why did those who were still free and outside the Camps, whose relatives and friends were being so dreadfully mutilated, not continuously protest? Ignorance of the truth, or fear of the consequences? In some cases, yes. But not always. Sometimes they did know, and they were not deterred by fear because they were abroad and safe. Then why did they, or *we*, not protest? The answer seems to be because they, and we, simply did not believe that it was possible for human beings to be so inhuman. And as for the people themselves, the victims, in prospect they too shared some of this unbelief and went, as it were, "like sheep to the slaughter", unprotesting until it was too late.

As Herbert Butterfield, the Oxford historian, observed: "We create tragedy after tragedy for ourselves by a lazy unexamined doctrine of man which is current amongst us and which history does not support . . . . Those who do not believe the doctrine of the Fall can hardly deny that human history has always been history under the terms and conditions of the Fall".

We suffer from a unique form of sickness which is not to be observed in nature apart from man. This sickness does two things: it makes us capable of truly incredible wickedness totally foreign to the animal world; and it puts blinkers on us which make us believe we are suffering from no such disease.

This disease is universal in man, and we all grow up to display it unfailingly. If one asks, "How soon is the delightful illusion of childhood innocence lost", one can only say that man sins just as soon as he can! Thousands of years of increasingly complex civilization have not really changed the picture. We are still as sick as our first ancestors were – Cain was a murderer. All we have done is to arm our wickedness with superior weapons of destruction. The disease lives on in the earth because man himself is the disease.

An individual, acting in defiance of society is bad enough. But there is probably nothing so wicked as a crowd acting in unison under the dictates of their lower nature. A culture may be so disrupted that a whole society goes bad. Authority is everywhere undermined to such an extent that lawlessness, destruction, violence, rape, murder, theft, and cruelty know no effective curbs and chaos results. This may happen in any social grouping such as a crowd. When a crowd throws off all recognition of established authority its mood changes rapidly from bad to worse, no longer constrained towards any good, but self-reinforced and self-reinforcing towards wickedness. Human behaviour becomes "liberated" *and equated with sin*. People are swept by the compulsive mood of the crowd, and individuals find themselves suddenly free to express the very worst side of their nature – often to their own genuine amazement in retrospect.

The roar of unified voices bent on evil is terrifying. There is something demonic about it. Crowds become vicious in ways totally foreign to the behaviour of the individuals who make up the crowd. Men in groups will become vicious murderers and violent in the extreme, even the gentlest of them. And history shows, sadly, that in times of great violence (as in the French Revolution) women are equally capable of cruelty. Even in watching violent sports, this unexpected side of woman's nature may be suddenly revealed. Afterwards, the individual may sort himself or herself out and ask in amazement, "What got into me?" Nothing got in. It is not what gets in at all but what comes out that reveals the truth of human nature, even as Christ said it would (Matt.15:18,19).

Man's dominion and government over the world has been a disaster. We do indeed seem to be very near to the end of the human experiment. Only a divine intervention can salvage what is left. The whole of human society is close to moral bankruptcy and technologically the resources of the world (air, water, minerals) are either almost exhausted or so hopelessly contaminated as to be no longer able to support life. Even outer space has become a junkyard.

What has happened is that man has become the arch-destroyer of his own appointed kingdom, turning a Garden of Eden which was filled to overflowing with everything good into a wilderness filled with the debris of our own folly and greed.

Some have even gone so far as to say that every desert area in the world is due to man's abuse of the land, and they base their arguments, as W. C. Lowdermilk has done, on the fact that most deserts are dotted with the remains of cities now buried under drifting sand. Other deserts are being attributed to a spin-off of man's unwise use of water resources which once formed part of a smoothly working ecology. In one area, Sir Daniel Hoare refers to a desert of over 40,000 acres in Africa which began with water running off a barn roof that was just allowed to carve a small channel which grew and grew because the farmer was too lazy to do anything about it. One native observer remarked, "Just one damn trickle forty years ago . . . . and now a third of the country gone".

Oliver Pearson says that man's impact on the environment has become so great that it is "probably greater than that of all other mammals combined. For many years man has been drawing on the earth's capital to support his high living; most other animals live frugally within the earth's income".

Andrew Ivy recently pointed out that "soil erosion and depletion caused the transformation of garden spots into deserts in Greece, Syria, Northern Italy, Mesopotamia, and the Uplands of China. We hear of dust storms in the Volga Valley, in South Africa, Australia, and the United States, the breadbaskets of the world". He might have added Canada to this.

Laura Thompson observed, "Man is not only a major factor in the web of life; he is the *only* agent whereby a conservation programme for a local area may be actively implemented". He alone is responsible for the upset; he alone can correct it. The trouble is man is literally too wicked to engage himself in any corrective process which requires any significant personal sacrifice. A. J. Carlson, with grave humour, wrote, "In the face of this can we claim the name Homo *sapiens*", man, the wise one? The one creature who is pleased to call himself such must seem pretty foolish to all other creatures, if they are able to judge him.

Man is not merely a disturber of nature because of ignorance. He is deliberately destructive because of some strange preference for destructiveness – even in childhood. Something is wrong with his nature. The beastliness of man is not of the beast. It was Alfred Lord Tennyson who in his poem *In Memoriam* (published ten years before Darwin's *Origin of Species*), coined the famous phrase "nature red in tooth and claw". In recent years a substantial number of books have been written which show clearly that nature is *not* red in tooth and claw, that animals are *not* aggressive towards each other in the sense that man is towards his own kind, that there is *no* vindictiveness or vandalism* in animal society, but that – as Prince Petr Kropotkin, after years of active research in the wild, put it – nature is characterized by a spirit of "mutual aid"! It is clear that the wickedness of man is not because there is something animal in his nature but because his nature is fallen.

When famous men like Professor George Gaylord Simpson and Will Durant persist in their defence of human evolution by arguing that the evidence shows that "man has risen – not fallen", they are talking unbelievable nonsense. History screams a negative.

---

* Some would argue that foxes are vandals when they kill hens indiscrimately, and so likewise when wolves kill sheep. The answer to this probably lies in the fact that the domestication of hens and sheep has destroyed their natural behaviour pattern under attack so that the predator has his natural instincts confused. Foxes do not do to wild fowl what they do to hens; nor wolves to wild sheep or goats what they do to domestic ones. The behaviour of the predator and the behaviour of the prey were balanced in nature, and man has upset the balance. If man had domesticated both predators as completely as he has domesticated both prey, perhaps this disruption would not be exhibited.

2. UNFALLEN MAN.

As evolution cannot account for fallen man, it cannot account for *unfallen* Man either. But where are we to observe unfallen Man that we can speak so confidently of what evolution cannot thus do? We find unfallen Man in the person of Jesus Christ.

Here was true Man, with a magnificent beauty of bodily form that made even those hired to arrest Him fall back when He stepped forward to identify himself (John 18:6), and an unutterable beauty of personality that was flawless though under constant provocation by his enemies.

Lord Acton, in a letter to Bishop Mandell Creighton in 1887, wrote, "power corrupts, and absolute power corrupts absolutely". If you wish to prove the corruption of human nature, give a man power over his fellowmen and the means to exercise it. The more the power, the more certain will be the display of corruption. But here was One who, although He had been given all power in heaven and earth and who although He could do whatsoever He wished consonant with the purity of his nature, nevertheless remained throughout a life of continuous challenge utterly uncorrupted.

The evidence of his power is everywhere to be found in the Gospels, but in no sense do they appear to the reader as examples of what we would view today as showmanship. In some strange way we expect them, for they are completely in keeping with everything else He was and did. Probably never before or since has a nation been no nearly rid of human sickness by the power of one man to command the source of it – whether the source was sin or demonic.

When we observe closely how He dealt with his challengers we can only stand back in amazement at his calm wisdom. One day the scribes and Pharisees, hoping to trap Him into making a statement publicly with which they could accuse Him of treason, asked Him whether it was proper to pay tribute to Caesar or not (Mark 12:13, 14). If

He replied, "No, it is not proper", the people would have cheered Him but his statement would have at once been reported to the authorities. If He had replied, "Yes, it is proper", the people would have turned against Him immediately, and the scribes and Pharisees would again have been the winners.

What did He do? He asked them to show Him a coin. The question arises why did He not have a coin of his own, since his little group had a treasurer. Perhaps He had a purpose in not appealing to the treasurer, who was Judas Iscariot. But the fact is that there were at least two kinds of coinage circulating in Palestine. The Romans allowed the Jews to mint coins of their own because they did not want to use the Emperor's coinage in their temple services – hence the existence of money-changers right in the temple precincts.

However, the scribes and Pharisees, bowing to their authority, preferred to use Roman coins for all commercial intercourse: and so the Lord turned to them and said, "Show Me a coin". It seems highly unlikely that they were aware of what He was doing, but when they showed Him one of their *Roman* coins, He held it up for everyone to see plainly and said, "Whose image and whose superscription does it bear?" To their shame the scribes and Pharisees had to say, "Caesar's". And they were trapped themselves, for by their very possession of these coins they were really strengthening the hold of the Romans on Palestine. Then He said, "Render unto Caesar the things that are Caesar's, and unto God the things that are God's". In other words, if you are going to use Caesar's coins, you must serve Caesar.

On this occasion several other challenging questions were presented to Him and He answered them all with equal ease and effectiveness. So much so, in fact, that some of the scribes themselves admitted defeat, saying, "Master, thou hast well said" (Luke 20:39), and after that they dared not ask Him any more questions.

Perhaps no one incident in the Lord's life displays his

extraordinary wisdom and gentleness than upon the occasion of his dealing with the woman taken in adultery. The story is given in John 8:1-11. This is a passage which many scholars today believe may not have belonged to John's Gospel in the original, because some of their favourite manuscripts from ancient times have omitted it. It should be noted, however, that the omission may have resulted from the fears of some copyists that in the story the Lord was really condoning adultery, and so they quietly deleted it. To my mind, the Lord was not condoning adultery but He was judging one who was no greater a sinner than the man with whom she was caught "in the act" (verse 4) – who had not, be it noted, been brought to judgment with her. One wonders why . . . . Perhaps the woman was more sinned against than sinning.

The law required that an adulteress be stoned, so the Pharisees brought this woman and flung her down in front of the Lord while all the people stood around. They brought the charge against the woman, pointed to what the law said must be done, and then posed their question, "But what sayest thou?" Notice the "But"!

If the Lord should say, "She must be set free as an act of mercy", the Pharisees could repudiate Him publicly for disregarding the Law of Moses. If He had said, "She must be stoned", it could only seem to the crowd around that He was merciless – righteous perhaps, but merciless.

So Jesus stooped down and wrote something with his finger in the dust which collected in the broad expanse of Solomon's Porch where these events evidently took place. He seemed to be ignoring them. Not unnaturally they were annoyed and persisted in asking the same question.

Jesus straightened up just long enough to say, "He that is without sin among you, let him first cast a stone at her".

It is rather widely agreed that the chief accuser had the responsibility of casting the first stone. This is a tribute in a sense to Jewish wisdom, for many who make accusations would not have the courage to do so – or perhaps the gall –

if they knew they were personally responsible for initiating the actual punishment itself.

Clearly the result of this simple invitation set them all thinking deeply, reconsidering their position. They started to leave one by one, as unobstrusively as possible, from the oldest of them down to the youngest, until they had all gone. And then Jesus straightened up and seeing that none were left, said to the woman, "Where are those thine accusers? Hath no man condemned thee?" And she replied, "No man, Lord".

Whatever we may think about the Lord's personal judgment, there is no doubt that she was, before the law, without any accusers, and could not legally be condemned. It would seem that the circumstances of her being taken were rather special, and perhaps Jesus knew what those circumstances were. At any rate He said to her, "Neither do I condemn thee: go and sin no more".

The story has a ring of truth about it, and it once more displays the extraordinary skill and wisdom of this Man. A wiser than Solomon was here.

In all his relationships with friend or foe, He preserved the perfection of his own manhood. This perfection was also reflected in his relationships with his mother. He knew how to respond to her claims when she sought them appropriately (Luke 2:51); to resist them when they were sought inappropriately (Luke 8:20, 21); and to recognize them when they were appropriate but unsought (John 19:27).

Such a figure as appears before us in the Gospels is truly uninventable. The literary creation of a character so perfect as this would require even greater faith than simply to believe the record as it stands. He is altogether and absolutely unlike ourselves, and the fact is scarcely denied even by his worst enemies throughout history.

When He was brought to trial by those who could not endure the white light of his purity, all kinds of people were presented as witnesses against Him but their witness was uniformly contradictory until it became clear to everyone that these witnesses were false. But there were many whose witness to his total innocence was almost involuntary, sometimes taking only the form of silence. John 8:46 records that the Lord once asked his accusers, "Which of you convinceth me of sin?" And quite frankly not one of them could think of a word to say.

When Jesus had been arrested and brought before Pilate, Pilate's wife warned her husband, saying, "Have thou nothing to do with that just man" (Matt.27:19). Pilate himself three times officially declared that he could find no fault in Him (John 18:38; 19:4, 6). On the third occasion he tried to be even more emphatic and exclaimed, "I am innocent of the blood of this just person" (Matt.27:24).

Even Judas Iscariot who had betrayed Him, went back to the chief priests and elders and offered to return the money he had received for his betrayal saying, "I have sinned in that I have betrayed innocent blood" (Matt.27:4).

One of the crucified men sharing some of his physical torture, rebuked his fellow in crime for speaking abusively to the Lord who was crucified between them, saying, "Dost thou not fear God, seeing thou art in the same condemnation? And we indeed justly; for we receive the due reward of our deeds: but this man hath done nothing amiss". How did he know this? He knew because *everyone* knew . . . . .

The Roman centurion in charge of the crucifixion detail of troops, after observing the behaviour of the Lord on the cross for a while, and no doubt having been responsible for many such events, said when Jesus died, "Certainly this was a righteous man: truly this was the Son of God" (Matt.27:54 and Luke 23:47).

Paul, the intellectual among the apostles, said, "He *knew* no sin" (2 Cor.5:21); Peter, the activist, said, "He *did* no sin" (1 Pet.2:22); and John, who loved Him most tenderly, said, "*In Him* was no sin" (1 John 3:5).

Never was there such a testimony to the total innocence of a man. So overwhelming was this witness that in the end the Jewish authorities themselves admitted they had made a mistake. They assembled to discuss the situation after the crucifixion and said among themselves, "Command that the sepulchre be made sure until the third day, lest his disciples come by night and steal him away and say unto the people, He is risen from the dead: *and so the last error shall be worse than the first*". (Matt.27:64).

The Lord Jesus came to reveal God to man, and He came to reveal man to God. But He also came to reveal Man to man, and this He did in two distinctively different ways. In the first place, He showed what true Man could be, and should be. He came a light to light every man that is born into the world (John 1:9). He came as a standard of reference, a plumbline, as Amos (7:7, 8) says.

If we want to know what we *ought* to be, here is our image restored. If, on the other hand, we want to know what we are *capable of*, given opportunity – whether man is essentially good, whether man loves truth, whether man really wants righteousness and purity and unselfishness and absolute integrity of person – then here again we have the answer. The only perfect Man who ever lived was condemned to crucifixion, and He was condemned not for some crime or evil deed or even falsehood, but for simply telling the truth about Himself, namely, that He was God and man – both (Matt.26:63-66).

It has been universally admitted by advocates of man's evolutionary origin that he is nonetheless "the crown" of creation. It is a strange thing that the most wonderful representative of this creation was by man himself crowned not with gold but with *thorns*. Is not this man's judgment of *himself*? How has such an anomaly come about?

A native from the Yana tribe once located in California who came to be known as Ishi (his own word for *man*) and who was the last lone survivor of his people, was shown a *Passion Play* film. He was deeply moved by the story of the crucifixion and remarked that Jesus Christ must have been a very "bad man" to suffer such a fate.

The truth is precisely the opposite. In the first place, had He been a bad man, God would not and could not have laid upon Him *our* sins: He would never have been acceptable to God as the sacrificial Lamb. In the second place, He *would*, on the contrary, have been acceptable to the world. But the world rejected Him and crucified Him not because He was a bad man but because He was a *perfect* Man! So it was only his perfection that made Him an acceptable sacrifice from God's point of view and an unacceptable person from the world's point of view. His very goodness, not his badness, was the reason He was condemned to death by us men. *We were* **all** *involved in that trial.*

Thus in reality the trial of Jesus Christ was not the trial of Jesus Christ at all, but the trial of fallen man. It was not He who was on trial, but man. And the outcome was not *his* condemnation but *ours*.

If it is not possible to build a bridge between animals and fallen man, what bridge shall be built between fallen man and the Lord Jesus Christ? If both are truly human, which they certainly are, how do we make a bridge between such purity and such utter wickedness? Is such a bridge possible?

The answer is, Yes! And to build this bridge we go back to the *first* Adam as created, and from there to the same first Adam as fallen. The Fall is the arch of this bridge. In that one act the First Adam (who is truly represented by the Last Adam) became also the fallen "Adam" who is represented by the human race as exhibited in the whole of human history. The potential, locked up in the newly created being called the First Adam was capable of giving rise to man as we see him displayed in the tragedy of history, and to the glory of God as seen in the face of Christ Jesus. What a creature this was who carried the potential for both the corruption of our nature and the perfection of His! Evolution is quite incapable of accounting for Him.

CONCLUSION.

Both men called Adam were immortals. Both men called Adam came into the world by a miracle: the one by creation, the other by virgin birth. Both men died – but neither need have done so, ever. Evolution cannot account for either of them.

However, the Lord's people are not being shown *why* it is so damaging to the Christian Faith to allow the evolution of man's body. The truth is that to do so is to divorce the Incarnation from its *redemptive* purpose and to reduce the life and death of Jesus Christ to one of tragedy rather than triumph. His virgin birth and his bodily resurrection become meaningless, since there is no rational necessity for either of them.

The suicidal wickedness of fallen man and the sacrificial splendour of Jesus Christ can only be accounted for by assuming a strictly historical basis for the appearance on earth of the First and the Last Adam precisely as they are set forth in the Bible. Neither their origins nor their deaths can be accounted for in ordinary biological terms, for neither were "natural".

May I conclude this chapter by saying that it was one of the most difficult to write in the whole book. It was completely re-written so many times that I despaired of ever getting it written at all, and at one point decided to omit it altogether. But it had to be done. Why, then, was it so difficult?

Because the immensity of fallen man's wickedness is beyond comprehension and because the beauty of unfallen Man's character is beyond description. That's why. Who can possibly be sufficient for *either* task.

Chapter 14

THE DYING OF MAN

**Man lives two lives and dies two deaths.
Spiritually, he commits suicide: physically, he is executed.**

The subject of *death* is an enormous one, and the literature is huge. Some thirty or forty years ago, I remember a scientific paper which opened by saying that over 600 books had already been published on the matter and at that time there was not even a glimmer of understanding of the cause of death except where there is accident or disease or predation to account for it. It is still, today, widely held that no one dies merely from the weight of years.

In man the problem is greatly compounded by the fact that whereas animals experience only physical death, man experiences two deaths – one spiritual and the other physical. As we live two kinds of life so we experience two kinds of dying.

Theologically, these two deaths can both be characterized by the single word *separation*. Physical death involves the separation of the spirit from the body, and spiritual death involves the separation of the spirit from God. In each case a 'termination' is reached which only God can reverse. There is the termination of physical life for which the only remedy is the redemption of the body (Rom.8:23), and there is the termination of spiritual life for which the only remedy is the regeneration of the spirit (John 3:3).

But this view of physical death is a gross over-simplification. In the first place, it is now recognized that death can be seen either as an *event* or as a *process*. From a legal and medical point of view it is an event, and the time of its occurrence can usually be stated. From a physiological point of view it is actually a process (as we have already noted in Chapter 6), which is going on throughout life and begins the moment we are born or even, perhaps, the moment we are conceived. What happens a few days after death is a further process of disintegration that is merely an acceleration of what has been proceeding since day one. The human body is corrupted from the very first, and this acceleration in the grave is only the last act in the play.

Nevertheless, we know that it need not be so, for there was one truly human body that never saw corruption either in life (1 Pet. 1:18, 19) or in death (Acts 13:37), though the burial conditions were not unlike those of Lazarus whose body did indeed see corruption (John 11:39).

Now the *immediate* cause of man's physical death can be a host of different things: starvation, disease, the accumulation of DNA replication errors, poisoning, suicide, fright, cold, heat, an accident, excessive joy, even laughter – and even hiccups!

The medical people always like to be able to establish the particular cause of death in each case if possible and the number of occasions upon which the death of an individual *seems* simply to have been natural and without any discoverable cause other than "old age" is remarkably few if there are any at all.*

Yet (as we have already noted) man does not seem to die a 'natural' death under *any* circumstances because

---

* The truth of this observation is documented n *The Seed of the Woman*, pages 3-15.

Scripture assures us that death was introduced into human experience as the result of an act of disobedience that involved eating a forbidden food. Unlike ordinary food poisoning, the end result in this instance was a fatal damage acquired during Adam's life time and, contrary to the normal rules of inheritance, a damage that was passed on to all of his naturally-born descendants. Thus by sin death entered into the world and as a consequence the human race was "un-naturally" mortalized.

Death for man is therefore the consequence OR penalty of disobedience. The decision as to which of these two alternatives is the correct one is still a matter of theological debate. We cannot be sure from Genesis 2:17 whether the Lord is saying to Adam that disobedience would bring death as a punishment or merely as a natural consequence. Perhaps it was both.

It is important to add that from one point of view, physical death was also a remedy, an act of mercy. As Methodius (d. c. 311) observed, and much later Francois Turretin (1623-1687), we must be rid of this defective body in order to be freed from the root of sin and hence of one basic cause of our fallen nature. Paul writes that the law was ineffective because of the "weakness of the flesh" (Rom.8:3), and the Lord excused the failure of his friends to watch with Him in a critical hour of his suffering, on the same grounds (Matt.26:41). We cannot do without the body, but there are times when we wish we could, because it is the source of a great deal of our spiritual failure. Freedom from a disposition towards sin hinges upon freedom from this "body of sin", a freedom which physical death guarantees for the redeemed.

It appears from the experimental evidence that animals have what is commonly called a "spanned" life, that is to say, a more or less predetermined span of life which is characteristic for each species. This is assumed to be necessary in order to prevent over-population by any one species.

When over-population threatens in nature, there are various compensating mechanisms to deal with the situation. Some species raise smaller broods; some develop wings (aphids for example) which enable them to leave the area; some commit mass suicide (lemmings, for example); some by their increase in numbers encourage more predators from contiguous areas who multiply in the presence of plentiful game and so reduce the larder until they, too, are reduced and things reach a balance once more (wolves and deer, for example). Some (like elephants) reduce their number not by smaller litters but by lengthening the gestation period very substantially. Thus the world's animal population remains in a remarkable state of balance between food resources and numbers to be fed – except, of course, where man gets into the act.

But if unfallen man was commanded to multiply and fill the earth (Gen.1:28), would he have lived on and on, multiplying indefinitely? What mechanism would have prevented *his* over-populating the world? Would not natural death have to be ordained for him also, as a safety device to prevent over-population by the human species? If it was, then natural death must have been ordained before sin had entered, a supposition that contradicts what Romans 5:12 says. The answer to this question is, No.

What *was* provided for an *unfallen* race as a safety device was not natural death but translation and transformation to a higher order of life that would not leave man disembodied but with a body which, like the Lord's resurrected body, does not occupy *space* at all. There was to have been no death, but only "graduation". In the divine plan it was man's destiny that he should never taste of death, though provision was made for him should his freedom abort the plan.

We are, I think, to view the Mount of Transfiguration experience of the only Man who never need have died, as providing us with a model to show what kind of transformation would have awaited us also if we had never sinned.

But there is this difference. For our sake, the Lord Jesus Christ came back down again and set his face to go up to Jerusalem and to death – "instead of the joy" that had been set before Him (Heb.12:2).*

Such a race, so transformed to the kind of physical existence that characterized the Lord's resurrected body† (of which we shall speak in Chapter 16) would in no way have over-populated the world! Like the angels, it seems we shall occupy *position* but not *space*. The problem of overcrowding of the world by immortals would therefore never have developed. For, as soon as each individual was made mature by the things which he experienced in this "time and space" existence, translation would have removed him to a higher form of existence in which time and space is of no significance. But as things are, death is necessary because no such transfiguration is in view for those not made perfect (i.e., mature) either by life as the Lord Jesus was (Heb.5:8) or by imputation as his people are (Heb.10:14). To be "made perfect" can be applied to the Lord only in the sense that whereas He was born *innocent*, He achieved absolute *virtue* by the experience of life. We never thus achieve absolute virtue. This is what it meant for the Lord Jesus to be "made perfect". It implies no imperfection at any time, but rather the purity of childhood which He turned into the positive virtue of perfect manhood.

Man does indeed die two deaths, but it is chiefly with his physical death that we are here concerned. Because his body was not designed for dying, and because the dissolution of the fundamental union between spirit and body is

---

* The Greek word *anti* is translated "for" in this passage in the King James Version, a small word which in the English idiom of that day meant "in exchange for" rather than "because of". According to Dana and Mantey, *A Manual Grammar of the Greek New Testament*, sect. 107, the normal meaning of *anti* at this period was "instead of".

† Shortly after his resurrection, the Lord's body was in some way transformed, since Mark 16:12 tells us that the disciples did not see Him as Mary had seen Him but "in another form" (Greek: *en hetero morphe*).

effectively the suspension of the wholeness of man, man's attitude towards death must clearly have a dimension to it that in no way troubles the animal world. For man, death is a terrible thing, a disaster to his very being: and he lives in fear of it for almost the whole of his life.

For animals, death appears to be a *natural* thing, whereas for man it is a wholly *un-natural* thing. When sin entered into human experience, death entered with it as something entirely foreign to him. Indeed, as Martyn Lloyd-Jones observed, it would be quite as proper, if not more proper, to render the word 'entered' as *invaded*. For as he points out, this is what the strong verb in the Greek really signifies.

In man, body and soul are so profoundly interpenetrating that the thought of separation is, as Thomas Aquinas put it, "utterly abhorrent". James Orr states the matter thus: death is "the violent rupture, or separation or tearing asunder, so to speak, of the two parts of his nature which in the Creator's design were never intended to be sundered . . . ." The immortality man was designed to enjoy was to be an immortality in which the body played an essential part. True immortality is not merely immortality of the spirit but embraces the resurrection of the body to make the spirit's immortality meaningful by providing the immortal spirit with a requisite and proper vehicle for expression.

James Denney spoke with perceptiveness when he wrote: "That which would be merely physical in the lower animals is *not* merely physical in man". While the consenting voice of science seems to say that the life principle in man's body is not different from that in any other animal, the Scriptures say that it is. It is different because man's body is a house designed for a spirit that has a totally different destiny, a destiny for which the body is essential. At the time of its dying, an animal's body has served its purpose and is laid to rest permanently, whereas the human body has only just *begun* to fulfil its purposes. Compared with eternity, a life in time is a mere instant; whereas the human body has a timeless eternity ahead of it. Its significance is entirely different.

Science asserts that man must die because all his antecedents in the animal world die. But this is simply not the case. In the first place, unicellular animals which according to evolution* must have been among his antecedents at the very beginning, were almost certainly not mortal. Thus if evolution is true the line of man actually began as an immortal one, not a mortal one! These organisms can be killed but they are not inherently subject to death at all. It would not be true, then, that death in the animal world necessarily lies at the root of the death of man.

In the second place, the assertion assumes lineal descent of man from one of the primate species, and for this there is no proof whatever. The only evidence is entirely circumstantial and has a certain weight only if we deny that the divine Designer might have used a similar pattern to produce a body by creation which was going to operate under similar conditions of physical existence. In less public enclaves of scientific discussion the absence of any such "proof" is now and then frankly admitted.

The evolution of man is an article of faith, not a scientifically proven fact, even though books written for public consumption feel the assertion can and must be made as though it were. In the nature of the case, it is impossible to prove genetically that there is any relationship between a fossil that looks like a man and the man that it looks like. It cannot be proved; it can only be argued as plausible. It has been said that "all fossils are foundlings", and establishing actual relationships with certainty is at present quite impossible unless there is some other kind of evidence – such as written documents, or an inscription on a tombstone, for instance. "Blood" relationships cannot be established

* After all, one of the most popular evolutionary slogans is the "amoeba to man" concept.

from bones that are completely fossilized.

Animal death is a mere termination of that particular animal. Human death is by no means a mere termination. It is neither termination of the individual's body nor termination of the individual's spirit. It is a disruption, but it is not a termination. It is quite natural in the one: quite unnatural in the other. It is a mistake to speak of any human death as natural. Common law in almost every society demands that the *cause* of death be established if possible, a fact which virtually denies that human death is ever natural. Indeed, natural death has been termed a "legal fiction".

The death of man is as much a spiritual event as a physical one because it is the result of a spiritual judgment. By contrast, the death of an animal is not a judgment at all but is according to the divine plan for the well-being of the animal world. In man it is a judgment because the consequences of unlimited physical life for a sinful creature were unthinkable. For animals it is a wise provision because of the necessity of avoiding unrestrained multiplication and over-crowding. If fallen man had had no such limitations placed on the length of his life, the accumulated experiences of wickedness carried on century after century could only have led to an appalling reinforcement of the corruption of his nature. It was indeed by reason of great longevity that the world had become so corrupt that the Flood was brought upon man to put an end to it all. Thereafter, a limitation of 120 years at that time (Gen.6:3) was imposed upon the normal life span, instead of the previous many centuries. In the same way, and for the same reason, it had been necessary to prevent man's access to the Tree of Life in the Garden of Eden, lest he should eat of it and recover the physical immortality with which he had been endowed in his unfallen state.

For this reason, as though the alternative was too awful to put into words, Genesis 3:22 appears in Scripture as one of the few unfinished sentences. It sometimes strikes me as interesting that my old Hebrew professor, Dr. T. J. Meek who was responsible for the Revised Standard Version of Genesis, indicated this unfinished-ness by the use of a long dash instead of a period. Yet Dr. Meek himself had no faith whatsoever in the Bible or its message, while many more recent translators who do have such a faith have not seen the significance of the unfinished structure of this sentence.

To account for this difference between the *death* of man and animals, we have to consider the corollary, the difference between the *life* of man and animals. Adam's death was no more due to his having inherited animal life than animal death is due to their being involved in Adam's sin. We shall never find the real meaning of life without also recognizing the real meaning of death. Man lives under sentence of death because of his sinfulness, and thus really lives out his life under judgment for a capital offence in which his body is also standing trial. Both spirit and body are under sentence of death. He thus experiences dying for a reason quite inapplicable to all other creatures not only by reason of its dual nature but by reason of its cause.

It is also important to realize that if Adam had been obedient, he would not have been *rewarded* with immortality as though it were a crown to be added to his stature. *Obedience would merely have ensured the preservation of what he already had.* When Adam sinned, he did not shorten his life: he introduced into it a tragic element, death, which till then was completely foreign to it.

What was thus imposed as a penalty was not a shortening of life so that he died prematurely. What was introduced was *death*, an entirely new and undesigned phenomenon. Immortality was never promised as a reward since he already enjoyed it, but loss of it was indeed threatened as a punishment. Retention of it was a reward only in a very special sense, but it was the *retention* not the acquisition of immortality that was the reward of obedience.

Thus man, unlike the animals, does not simply come to

an end. His death is by appointment, an appointment with the Judge which is followed by the passing of sentence on his life (Heb.9:27). Death is, therefore, tantamount to a summons to the Courthouse, and the summons carries with it the certainty that, apart from saving faith, the judgment can only be "guilty" – even before the trial begins. He thus lives all his life under the shadow of a warrant of death and when the time comes he is executed. This is what makes death so terrible, apart from redemption, and so utterly different for man from the death of every other creature.

Now we have spoken of two lives and two deaths in man's case, and it is necessary to say a few words about spiritual death. While physically man is "put to death", spiritually man actually "chooses" to die. This exercise of choice in the matter of spiritual death means that effectively man commits spiritual suicide. By which I mean that we sin and die spiritually because we *want* to. We don't like the penalty, but we do want to sin. Such is the nature of *human* nature. And as already noted, we sin "as soon as we can".

We go willingly along this route: we do not have to be persuaded. We are invited to choose spiritual life but we choose spiritual death, unfailingly and universally. This is the natural inclination of the natural man even though at first we are aware that it is not the way to go and our conscience troubles us.

When we commit sin we are acting freely. Herein lies whatever freedom was left to us after the Fall: *freedom to sin*. Like a free-falling parachutist, we are not aware of our actual bondage unless we suddenly try to go the other way. We do not sin with any compulsion from without; we sin because of an inner drive, a drive which is suicidal with respect to our spiritual life.

By contrast, in the matter of physical death, the case is

entirely different. No man in health *wants* to die. While our spiritual life is willingly surrendered, our physical life is not at all willingly surrendered. We are executed. We are unwilling *victims*. Death happens to us, overtakes us. We may hasten it by bad habits but we do not deliberately adopt those bad habits in order to hasten it. We would prefer to enjoy the bad habits without the hastening....

There is always, of course, in the matter of physical death viewed merely in its physiological aspects, a certain parallelism between the decease of animals and the decease of man, but this is because man was designed to function in the same world. Yet for all this parallelism there is a fundamental difference nevertheless, because man was not designed for death. Though he no longer enjoys the kind of physical life which God intended for him, the *potential* for it still remains. By the experience of new birth and redemption that potential will be realized once again in due course.

In this, therefore, there is clearly a difference between animal and human death. Man knows he must die and hates the thought of it. At the last moment, it would seem to us that an animal would "hate" the prospect of it, but they don't *know* that they must die. They have no knowledge that there is a spanned allotment of days, whereas man thinks otherwise of himself and seeks to postpone death if at all possible. And not infrequently the medical profession imposes prolongation upon him under conditions which would sometimes even be viewed as cruel if it were applied to animals.

Why do we have such a horror of dying personally and extend this horror to the dying of others even when it would clearly be a merciful thing to allow them the freedom to do so? There was a time when the medical people sought to preserve or improve *health* .... not simply to prolong life per se. The goal was to add life to years, not merely years to life.

Superficially, the dying of an animal and the dying of a

man might seem to be quite similar, but there is a tremendous difference in the very nature, the very *essence*, of a life lived under the conditions of potential immortality as opposed to life lived under the terms of a planned limitation that the animal cannot possibly be aware of.

For man, the imposition of death lies heavily upon his inner consciousness as a dark shadow. Animals know nothing of this "shadow of death", so that there is a difference between the death of fallen man and unfallen animals despite the fact that it is for both a termination of something.

Watching a herd of wild creatures being attacked by a predator, one becomes aware that as soon as one of their number is brought down, the frantic rush to escape seems to be treated at once rather as a form of excitement, almost more of a stimulus than a terror. The rest of the herd stops running and all resume their browsing or their play. That one of their company has suffered a violent death appears to have little or no upsetting effect whatever, and under normal circumstances both prey and predator take no further interest in each other. This is not man's attitude towards death at all. While physiologically the two deaths may be described in the same terms as to their effect viewed externally, they are clearly quite different as experienced internally.

It might be argued that the machinery of the body, like all other machinery, is bound to be subject to failure in the end. Death would therefore be inevitable for man and animals alike. We know now that this is not the case. It is not true of unicellular creatures like the amoeba. The microscopic size of these creatures has little or no importance since *size* is irrelevant to life. To the individual amoeba a lifetime is a lifetime, and its highly active awareness of its situation, so ably recorded by H. S. Jennings in 1910, is just as real an awareness for these little creatures as ours is for us. Their cup of awareness is full. The whole gamut of its responses to life's challenges are as serious to it as are children's fears and hopes and disappointments to them. We as adults forget how real a child's disappointments are since they seem by comparison with ours so inconsequential. But, I repeat, size has really nothing to do with the situation.

It is clear, therefore, that God can and has designed creatures which are not subject to natural death, though these creatures experience life in the fullest sense of the term for them. It would be a great mistake to suppose that these microscopic animals are not complex simply because they are so small. All living systems are unbeliveably complex, not only in structure but in behaviour as well. Even individual body cells taken apart can reassemble themselves and carry on! *Nothing* that lives is "simple".

There is no reason why the machinery of the human body should not have been perfectly designed to operate indefinitely and without failure. Indeed, we know that it can operate perfectly because the world has witessed One such perfect organism which was truly human – the body of the Lord Jesus Christ. Our redemption hinges upon the perfect operation of that body, since if it had been imperfect as our fallen bodies now are, it would have been destined to die anyway and the sacrifice of that body could have been no more than a premature, not a vicarious death.

Man finds it difficult to think of himself apart from his body. The death of the body is therefore felt as a threat to personal continuance. It is both a rending asunder and a potential termination of identity.

But God "hath set eternity in the heart of man" (Eccl. 3:11)* and throughout history man has borne witness to this deep conviction by caring for the dead in a way animals never do – by deliberately burying them and often trying to ensure their comfort in the world to come. It is almost universally agreed by anthropologists that wherever fossils

---

* The word rendered "the world" in the King James Version is the Hebrew word for "eternity".

are found buried with provision for a future life, no matter how simple and fragmentary the evidence is, those fossil remains belong within the human species.

Of course, where no such evidence exists the remains may still be truly human, for people are sometimes buried by accident. But the presence of pottery vessels or figurines or food of any kind, or even burial in a foetal position, this is generally taken as evidence of conscious concern for a life hereafter. Our primitive contemporaries, who have in the past been viewed as "our contemporary ancestors", are even more likely to leave such evidence in the grave than we who consider ourselves much more advanced in our ideas. They more easily sacrifice valuables to this end.

No animal shows any of this kind of concern for its dead. Death is clearly a very different matter for man than it is for animals.

Evolution may very well provide a rationale for the death of animals, but in relation to the death of man, as we experience it in all its sadness or terror, evolution really has nothing to say. It is a different phenomenon, an unnatural one and therefore not accountable by derivation from death as animals experience it. It belongs in another category, and only revelation can shed any real light on its meaning for man.

While I greatly admire those who have so ably defended creation against evolution, I cannot help but feel that to do this by deliberately divorcing the issue from Christian Faith is to treat the case as though it were merely a matter of "scientific evidence". It would seem to be humanly wise, but I fear it is really a spiritual surrender to secularism.

The issue has to be fought on our grounds, not theirs. If it is won on their grounds and the teaching of creation is ever allowed, it will be a victory of the intellect but will have lost its spiritual significance entirely. The theory of creation can never be presented faithfully as an alternative to evolution by divorcing it from its spiritual implications.

Chapter 15

THE DYING OF JESUS CHRIST

Animals die – NATURALLY – by design,
Man now dies – UN-NATURALLY – by execution,
Jesus Christ died – SUPER-NATURALLY – by an act of will.

Man dies *two* deaths.

The Saviour of man must therefore also suffer *two* deaths, first by dying spiritually as man dies spiritually, and then by dying physiologically as man dies physiologically. For such a Saviour both deaths are substitutionary, unique as to their nature and cause, and unique as to their effect. If the actual nature and cause of either death was the same for the Saviour as it is for us *fallen* men, they cannot be substitutionary because in our case both kinds of death are proof of personal guilt.

Both dyings are acts of separation: the separation of the soul or spirit from the divine presence (which is spiritual death), and the separation of the spirit from the body (which is physical death).

Let us consider these two dyings of the Saviour in order that we may see how the Word of God makes them entirely distinct and opposite in both character and effect to the spiritual and physical dyings of man which have resulted from Adam's disobedience.

I. THE SPIRITUAL DYING OF JESUS CHRIST.

When man sins, he does so by choice and he thus commits spiritual suicide. It is an act of freedom: he elects to do it, and pays the price accordingly. "The soul that sinneth shall die"(Ezek.18:4,20). Thereafter he is, spiritually considered, a dead man: dead in trespasses and sins (Eph.2.1), dead towards God whom he must now face not as his heavenly Father but as his Judge. He no longer has the immediate access that he formerly had, and as he matures in his fallen state the sense of God's presence gradually declines, to be replaced by a surrogate god after his own image and compatible with his own nature because of his own making.

From the God and Father of our Lord Jesus Christ he has effectively cut himself off, separated himself – as Isaiah 59:2 puts it – until he becomes aware not only of his isolation but also of his comparative indifference to the loss. Paul describes this spiritual death as "eternal destruction from the presence of the Lord" (2 Thess.1:9, RSV).

But it may be asked, When did this kind of dying ever happen in the life of the Lord Jesus Christ? The answer is, On the cross in those three hours of darkness as indicated when He cried out in his extreme isolation and agony of soul, "My God, My God! Why hast Thou forsaken Me?" For in becoming an offering for our sins, He had suddenly experienced for the first time in the eternity of his being "destruction from the presence of the Lord", a destruction which for all He knew was final. It happened not by his choice (as it is with us) but by imposition when He was made sin, when He became guilty by imputation of all the horror and frightfulness of man's wickedness since history began with the murder of Abel.

When this judgment fell upon Him, it was as though the murder, the torture, the rape, the mutilation, the degradation, and the utter cruelty of man towards man, became in effect *his* responsibility. When all the diseased conditions of man due to sin were laid upon his soul, He assumed the responsibility for them, being afflicted with our afflictions. It was a cup He had anticipated with such horror in Gethsemane that He cried, "Father, if thou be willing, remove this cup from me: nevertheless, not my will but thy will be done" (Luke 22:42).

Herein we observe the first fundamental difference between his experience and ours. We become sinful by choice: He, entirely against his will, willing it only in the sense that He *surrendered* to his Father's will that He should assume the burden of it. Thus every reference to this experience of separation from the Father because of sin is described by the use of the 'passive voice' signifying not a doing of his own but something done to Him. It was imposed upon Him, not assumed by choice, only by resignation.

Isaiah 53:4, 5 and 8 hammers this home: "We esteemed Him stricken (passive), smitten of God (passive), and afflicted (passive). He was wounded (passive) for our transgressions, He was bruised (passive) for our iniquities . . . . He was cut off (passive) out of the land of the living: for the trangressions of my people was He stricken (passive)". In 2 Corinthians 5:21, "He *was made* a sin-offering for us", once again reflecting the same passivity.

And so it goes: always it is passive. In no way was his involvement in our sins something He sought as though it might be pleasurable, for pleasurable they can certainly be to us (Heb.11:25). Whereas man becomes engaged in sin with a kind of eagerness, the Lord faced the prospect with stark abhorrence. And when the pall of guilt fell suddenly upon his soul, He cried out in desolation, yet no longer to his Father but to a righteous God who now stood as his Judge. When the darkness of sin enveloped Him as He bare our sins in his own body on the tree (1 Pet.2:24), his cry of despair was a cry of one totally forsaken. He was indeed "cut off", though not for Himself (Dan.9:26), cut off out of the land of the *living* and counted among the spiritually dead.

We cannot really have the slightest conception of what this experience meant to One who was completely without sin. It was a sudden total defilement. We ourselves are already sinful and when we compound the evil by further sins we feel little shame because we are innured to its effects. To Him it was truly his first personal, and only personal, experience of guilt.

Now I know that any illustration from my own life will not really help here. Nevertheless I once experienced the impact of sudden contamination in a small way. At the time, I was comparatively "innocent" in the sense that any grown man may be innocent who has not yet had opportunity or occasion to be much else. I do not believe I had ever committed any "great" sin up to that time. My conscience was free, and inwardly I felt clean. I had been the Lord's child for some years and had an active ministry with young people which, I may add, continued for many years after this incident.

But as a consulting engineer I had been unduly charmed by a member of the administrative staff where I was working. One lunch time, I found myself alone with her in my part of the building and in a moment of idiocy I kissed her. I may say she was quite unresisting. It would, I'm sure, have been viewed as a very small thing by many of my friends, certainly by my worldly ones. But I immediately fled from my office and out of the building and walked quickly along back streets in an attempt to be alone. It was lunch hour, and I had to escape. I felt utterly, utterly sick and inwardly defiled. I was appalled at what I had done. This is simply how I *felt* about it. I felt diseased, contaminated, self-condemned.

I cannot put into words how the shame and sense of sinfulness flooded over my soul. It was springtime outside, but in my heart it was a cold, miserable gloomy day: the sun seemed to have gone in, the buildings looked drab and empty, the world was cloudy and hostile. I walked with my head on my chest and my spirit broken. I felt sick enough to vomit on the street. I was truly appalled and felt certain my future was ruined, my job in jeopardy, and every man must be ashamed for me. I felt a terrible need to tell someone what I had done, to confess and weep over it all. I cannot fully express today the darkness in my soul for the next hour or so until gradually my sense of the Lord's presence was recovered and I went back to my office a sober and a chastened man.

Somehow the Lord graciously covered it all. In a fallen world such an act would hardly cause a stir, I suppose – though for me it had been a devastating experience. It all happened so quickly, unplanned, unpremeditated, unexpected. It was like being plunged suddenly into icy cold water.

If a man, already fully aware of his own fallen nature, can be so devastated by such an act, what must have been the effect of the defilement of sin upon the Lord Jesus in view of his matchless perfection and purity and holiness? Any first scratch on a new car, any first mark on a new cabinet, any first water stain on a new wall or ceiling, is distressing in a special way. What would the first impact upon his soul be as sin was laid upon Him? When He suddenly became identified and responsible personally for all the sins of human history in their total appalling immensity, the effect must have been shattering.

I think that when we are told He bore our sins, we are to understand that He really in his own heart and mind became consciously guilty as though He had really committed these things. The effect of such a consciousness would be multiplied infinitely by the very sensitivity of his nature and his ability to identify so immediately with the effect of sin in the world around Him.

When the blow fell, his Father turned from his beloved Son and forsook Him in the darkness of his hell while the very light of day failed and plunged the world around Him into unnatural darkness. And for how long? For three hours? No, surely not, but for an experienced eternity.

Our clocks ticked on for three hours of course, but the passage of time for Him was swallowed up in the now-ness of the *intensity* of that suffering. All sense of the passage of time, all that might have signalled to his consciousness that the end was near, must have been eclipsed entirely. What was experienced began to occupy an eternity.

As Luther put it, with his characteristic bluntness, "Christ was accursed, and of all sinners the greatest. My sins caused Thee, dear Lord, to bear the wrath of God and become a curse, to taste the anguish of hell and to endure a bitter death .... Christ had to feel in his innocent tender heart God's wrath and judgment against sin and to taste for us eternal death and damnation, and in a word, to suffer all that a condemned sinner has deserved and must suffer forever".

Jonathan Edwards wrote about the matter thus: "Sin must be punished with an infinite punishment .... the majesty of God requires this vindication. It cannot properly be vindicated without it, neither can God be just to Himself without this vindication". Satisfaction for sin demands equal penalty, and not until that penalty had been paid in full did the Father turn his face to his beloved Son again.

It has been found that the Greek word which we recall so well in the translation "it is finished", was written across the bottom of Statements of Account in ancient Greece, and clearly meant PAID IN FULL! This cry, "Paid in full!", was therefore a cry of triumph, marking the end of his spiritual dying, an end which He could not possibly have foreseen so long as He was locked into the agony his suffering entailed. And because no end had been foreseeable, it had been effectively experienced as an eternity.*

Suddenly, it was all over. The sun broke through the gloom, the fellowship of the Father was his again; and the

---

*** How a mere three hours by our clocks could for Him be an eternity is explored in depth in the author's *Journey Out of Time*.**

"My God!" of that black night became the "My Father" of his last spoken words from the cross as He commended his spirit into his Father's hands.

Such was the spiritual death of the Lord Jesus Christ experienced *as* Man and *for* man, that the way might be opened for the forgiveness of man's sins. All that is asked of the sinner now is that he believe this and accept this spiritual death of the Saviour as a full perfect and sufficient sacrifice and satisfaction for his sins.

In this death, He had died for man's SINS. There yet remained a dying for man's SIN.*

II. THE PHYSICAL DYING OF JESUS CHRIST.

We have observed that when man dies spiritually his death is the direct result of an action deliberately undertaken. To act sinfully is pleasurable, at least in anticipation. Sometimes it seems almost as though this would be the definition of what is sinful – "that which gives pleasure"!

What ever the form our particular sins happen to take (and for each individual they differ according to personal preference!), sins are freely engaged in even though we may sometimes persuade ourselves that we are "doing it against our will". We all die spiritually – or better, we all *destroy ourselves* spiritually, by active choice. Spiritual death for man is indeed a form of suicide because it is voluntary.

---

* As man lives two lives, so he dies two deaths. In both realms, the physical and the spiritual, the effect of the Fall was to introduce *death*. SIN brings death of the body: SINS, the death of the spirit. SIN is now the cause of physical death, and it is the root of SINS of the spirit. SIN is to be *taken away* (John 1:29), *put away* (Heb.9:26), and *cleansed* (1 John 1:7), but not forgiven; whereas SINS are *forgiven*. Accordingly, on the Day of Atonement *two* goats were "sacrificed", the one as a SIN-offering (Lev.16:9), the other (the scape goat) for the SINS of the people.

In the theology of the Epistles, the distinction is constantly being assumed. John Calvin recognized this and commented on it succinctly (*Institutes*, Bk.2, ch.1, sect.5). Peter Lange in his *Commentary on Holy Scripture* (Romans Vol.X, Zondervan reprint, p.176, col.a) has a most useful excursus on the subject. Griffith Thomas points out that "The Bible distinguishes between SIN and SINS, the root and the fruit, the principle and the practice; and Article II of the *Thirty-Nine Articles* teaches that our Lord's atonement covers *both* of these" (*Principles of Theology*, Baker reprint, 1979, p.50).

On the other hand, the Lord Jesus Christ became involved in our sins with a perfect hatred, so fulfilling Psalm 139:22 in its deeper sense. For Him it as a far more dreadful involvement than the experience of physical death because spiritual death is a kind of continuous dying whereas physical death is normally a once-only experience.

For Him spiritual death was a totally unwilling involvement, something undertaken with horror. For Him it was an execution. But now He must also undertake to experience the "tasting" of *physical* death in order to complete his work.

Here we have a total reversal. For men, spiritual death is a form of *suicide* but physical death is clearly an *execution*. For the Lord spiritual death was clearly an *execution*, whereas his physical death was effectively a kind of *suicide*. Such is the contrasting position of the two Adams in their two kinds of dying. Let us look at the evidence for this statement.

First of all, we have to bear in mind that while, by reason of his Incarnation He had made Himself vulnerable and could therefore have been killed by accident or by human hands at any time had this been allowed of God (cf. Matt.4:6 and John 19:11), He had nevertheless by his supernatural conception escaped the mortogenic factor that we all inherit which brings us inevitably to the grave. He was, therefore, truly in the position that it was quite possible for Him never to have experienced death at all. After being transformed on the Mount of Transfiguration, He might have gone on into glory without ever passing through death and never have returned to his earthly life. Instead, He came down from the Mount, and did so specifically that He might taste of death.

But to what extent does the New Testament support the view that his physical death was wholly without compulsion, a dying of active choice and not merely a willingness to be put to death, a deliberate dismissal of life without any external or circumstantial constraint in the process?

The New Testament witnesses to this extraordinary fact in a number of remarkable ways that have largely remained unrecognized in commentaries of the past century or so, but which were clearly perceived by the earliest of the Church Fathers and by not a few of the earlier Reformers. Perhaps the reason for this comparative silence on the matter today is that it is difficult to state the case with clarity and precision without labouring the point, and to labour such a truth has something of impropriety about it. Perception of the truth has to be left largely to the reader.

There are three points of view from which his death can be considered: 1. as a HISTORICAL fact: 2. as a MORAL fact: and 3. as a THEOLOGICAL fact. These three conceptual points of view can be treated quite separately and are so indicated in Scripture.

1. The HISTORICAL View.

In the simplest terms, Jesus Christ was *crucified* and *slain*. It may seem rather unnecessary to emphasize the order of these two words, *crucified* and *slain*, since it seems so obvious that crucifixion was a mode of capital punishment. But in point of fact the phrase is rarely found in Scripture in that order, since in Jewish practice condemned men were *not* crucified and slain, but *slain first and then crucified afterwards*. Crucifixion was performed by the Jewish people for the sole purpose of *shaming the dead* and never as a mode of execution.

Crucifixion is generally considered to have been a Carthaginian invention, and was used there as a form of maximum punishment, inflicting death with the greatest possible cruelty. Not the least element of cruelty was the extraordinary time of survival on the cross before death overtook the victim and freed him from his agony. Both men and women have endured crucifixion for up to nine days before death has set them free. In some cases men so crucified have remained there alive long enough for birds to peck out their eyes; and they were of course totally unable to defend themselves. It was a truly awful form of execution.

The Romans had adopted this form of punishment for treasonable offences and for even lesser offences among commoners. But in Palestine they had been persuaded by the Jewish authorities to make a concession. It became permissible to kill the victim before the end of the Jewish day (which came at 6 p.m.) in order that the body might not be left on the cross beyond sunset. The Jewish people had long practiced crucifixion – or as the Old Testament has it, "hanging on a tree"; but the Mosaic Law had forbidden the retention of the body on the cross after sundown in order that the land be not defiled (Deut.21:22, 23). Burial of some kind was required before the day's end. As a consequence, the victim's death had to be ensured before sundown and this was carried out by the simple device of breaking the legs presumably to induce a form of suffocation, the whole weight of the body being thrown on the shoulders and chest.

However, the Jews themselves never once crucified men alive as far as we know. For them, it was never a form of execution, though it had been practiced for centuries all around them. Prior to the Roman occupation, it had always served the sole purpose of desecration of the *dead*. This fact is amply borne out by reference to such passages as the following: Josh.8:29; 10:26; 2 Sam.4:12; Gen.40:19 (from Egypt); and in the New Testament, Matt.23:34. The order of events is always the same: first slaying, and *then* crucifixion.

In their view, and according to Mosaic Law, any person thus "hung on a tree" was doubly cursed: cursed by society, and cursed by God (Deut.21:23). And in the Lord's case, there is no question that the Jewish authorities wished the common people to see that Jesus, by the very fact of his crucifixion, had been *accursed of God* – thus effectively undermining any claim He might have made as their Messiah. The very fact of crucifixion totally invalidated any such claim.

But there was another very important reason why crucifixion was necessary for the Lord, and why no other form of execution would have served God's purposes. He had to die TWO kinds of death, and his execution had therefore to be sufficiently prolonged that He could fulfil both while in this position of condemnation. Just as there had to be two goats on the Day of Atonement (Lev.16) in order that these two kinds of death might be accomplished – one the "sending away" to the wilderness of the scapegoat which foreshadowed the hours of darkness on the cross, the other the shedding of the blood of the second goat which foreshadowed His physical death on the cross – so there had to be *time* for the accomplishment of these things on Calvary: time for both kinds of dying of the Saviour.

In any one of the then current Jewish forms of capital punishment (strangling, beheading, stoning, etc.), only a *miracle* could have kept the Lord alive to perform these two functions of the Atonement, occupying as they did several hours. As it turned out, it was only by a *miracle* that He died when He did – and this solely because He was condemned to be *crucified.* No other setting for his dying could have accommodated these things which were essential to make his total sacrifice truly an offering for SINS and an offering for SIN; and in that order.

It was not, therefore, the crucifixion that really ended his life. He died ON the cross, but not FROM crucifixion. The cross was the occasion but not the cause of his dying. He was dead within less than six hours, a circumstance almost unheard of. The minimum time to death by crucifixion has been established by some writers of that time as about 32 hours, or over five times as long. It is therefore no wonder that the centurion in charge of the crucifixion detail was so amazed (Mark 15:39), and that Pilate also was incredulous (Mark 15:44) that He was so soon dead. Both were Romans: both probably had had considerable experience in such matters. To them it was a most exceptional circumstance.

2. The MORAL View.

How, then, does the New Testament describe this extraordinary situation? Partly, as rather frequently in Scripture, by the use of contradictory statements in order to arrest the attention of the dedicated reader.

Thus when Peter preached his first sermon dealing with the Lord's death, he accused his contemporaries of having by wicked hands crucified and slain the Lord (Acts 2:23). But when he preached his second sermon he again accused them of the same crime by saying "whom ye slew and hanged on a tree" (Acts 5:30) – this time, it will be noted, reversing the order. First they had crucified and slain the Lord: now he tells them that they slew and crucified the Lord.

In the first instance he was stating what was *historically* the order of events but in the second case he was giving the *moral* order of events, which was in fact much nearer the truth of the matter. For these same Pharisees had indeed already slain Him by the time they demanded his crucifixion. *They had hated Him and in their hatred had effectively murdered Him*, for hatred is murder (1 John 3:15).

In fact, crucifying Him was not, in their minds, to secure his death but rather to totally discredit Him. Simply to have had Him deliberately put to death by Pilate would have made Him a hero or a martyr. They may even have believed that if He did die on the cross He would indeed have been proved an imposter and thus their bringing Him to justice would turn out to their credit.

With respect to Peter's transposition of the words "crucified and slain", some versions have rendered this passage in a way which appears to contradict the sense of the King James Version. It is translated "whom ye slew by crucifying Him", thus reconciling the second sermon with the first one. It is a possible rendering. But it is actually contradicted by what we know about the Jewish attitude towards crucifixion. They did not slay by crucifying.

It will be noted that this reversal of order is also to be observed in Acts 10:39. Here Peter again contradicts his first sermon, placing death *before* crucifixion. Yet to place death before crucifixion is simply to affirm what is said elsewhere throughout Scripture of the order of events in such a case. In Peter's hearing, the Lord Himself had said that they themselves would be killed and crucified (Matt. 23:34). From all of which I think it is fair to say that Peter's perception of the matter had been sharpened as he had later reconsidered some of the very specific statements the Lord had made to the effect that no man was going to take his life from Him, but rather that He was going to lay it down of Himself.

And so we come to the third aspect of the crucifixion.

3. The THEOLOGICAL View.

Matthew, Mark, and Luke present their record of events with little or no theological comment. John's Gospel is quite different in this respect. It is in John's Gospel that we find the words of the Lord, "I lay down my life for the sheep . . . . Therefore doth my Father love me, because I lay down my life, that I might take it again. *No man taketh it from me, but I lay it down of myself*. I have power to lay it down and I have power to take it again. This commandment have I received of my Father" (John 10:15, 17, 18).

The Greek of these verses is so clear and straightforward and simple that it would make a beautiful exercise for any beginner studying the language. One of the most striking things about it is the repetition of the words, "I lay down my life". This phrase occurs in verses 15 and 17 and twice in verse 18. It is doubtful if the Lord could have placed greater emphasis upon anything He ever said to his disciples than He does by these simple words "*I lay it down of myself*".

As I read these words, I see the Lord trying to impress upon his disciples that in no way is his life going to be taken from Him. He is going to die but it is going to be his act. He will not die as other men have died under compulsion. Nor will He merely choose the time when He will allow other men to put Him to death. In the simplest possible terms, the very act of dying will be *his* choice, regardless of circumstances.

When John in his Gospel came to record the Lord's death, he used a word never elsewhere used in classical or biblical Greek for the death of any other man. In John 19:30 we have these words: "When Jesus therefore had received the vinegar, He said, It is finished, and he bowed his head and gave up the ghost".

The impression one has from this is that He yielded up his spirit under the pressure of circumstance, even as Ananias and Sapphira are both said to have yielded up their spirits. But the Greek word for "gave up" which John employs in this instance is not at all the normal word used for expiration. It is the word *paradidomi** which means *not to surrender* but to DISMISS. But the other Gospel writers, Matthew, Mark, and Luke, all employed words which were commonly used in Greek to describe man's passing, and indeed so did Peter in Acts. By contrast, Paul uses this word *paradidomi* on a number of highly significant occasions when speaking of the Lord's death, as will be seen by reference to Galations 2:20; Ephesians 4:19; 5:2 and 25.

When the Lord said He had power to lay down his life, He made his point clear by adding, "and I have power to take it again". It is obvious that while any man may commit suicide – i.e., has power to terminate his own life,

---

* **A full discussion of this point will be found in *Seed of the Woman*, p. 412 f.**

it is equally obvious that he does not have power to take it up again. The Lord had the same power both to lay down and to take up – or to put the matter slightly differently, He was entirely in charge of the process, *both ways*. He dismissed life by an act of will, and by an act of will later re-engaged it.

Thus in his exercise of absolute authority over his own life He did not give up his spirit in the sense that other men give up theirs. He deliberately dismissed it, and the transformation of his body from a living organism into a dead body was so immediate that the centurion was amazed.

No wonder, therefore, that it was a man in authority who could say to one under his command, "Go! and he goeth", who suddenly perceived that One far mightier than he had been able to say to his own spirit "Go!" and it obeyed immediately. Thus had this Man died, at his own command and in no sense in obedience to a summons from any other authority. He cried out in triumph, "It is paid in full!", commended his spirit into his Father's hands, and deliberately "blew out the candle". When the time came, it all happened in a matter of minutes.

He had thus accomplish the work his Father had given Him to do, partly in *an eternity*, and now finally in *time*.

Thus, *historically* considered, Jesus was crucified and slain. From the *moral* point of view, they slew Him by their hatred and sealed it by a crucifixion. *Theologically*, the cross was only a stage upon which the Lord Himself voluntarily became his own executioner. These aspects of the crucifixion which are so seldom spoken about in modern commentaries, have been recognized from the earliest times.

Tertullian (c. 160 - c. 215) wrote: "Christ when crucified spontaneously dismissed his spirit with a word, thus forestalling the office of the executioner". Two hundred years later, Methodius, Bishop of Olympas, observed: "Christ chose death to which He was not subject, that He might deliver them which were the bondslaves of death".

In 1886 Alfred Edersheim put the matter this way: "His death, his resurrection – let no one imagaine that it came from without! *It is his own act*. He has power in regard to both, and both are his own voluntary, sovereign, and divine act".

In 1895 James Denney wrote a little more extensively, saying, "If death was precisely the same problem for Christ that it is for us, then the New Testament way of speaking about his death is simply incomprehensible. If the first Christians had been of this mind, the phraseology we find in every page of Scripture could never have arisen. But they were not of this mind. They believed that Christ was sinless, and therefore that death, although included in his vocation, had a unique significance . . . . *his death is a solitary phenomenon*, the one thing of a kind in the universe – a sinless One, submitting to [I would have said *embracing*] the doom of sin. It was *his* death, certainly, for He had come to die; but it was *not* his, for He knew no sin; it was for us, and not for himself that He made death his own".

Fifty years later, John Murray underscored what we have been saying as follows: "(The death of Jesus) was unique because of the way in which He died. *No other died as He died*. How can this be? All others die because forces other than their own wrest life from them and sever the bond uniting body and spirit. Not so Jesus on the accursed tree. He was indeed crucified by others: He did not crucify Himself. But when He died, He dismissed his spirit, He laid down his life: He, in the exercise of his own agency and by the authority given Him, severed the bond."

The difference between his dying and our dying can be illustrated by a series of short antithetical statements, which can be documented from Scripture either in the actual wording or as clearly implied. They may be tabulated, though over-simplified, as follows:

We *are* subject to death. . . . . . . He *became* subject to death
We *are* humbled by death . . . . He *humbled* Himself
We, like Paul, are "offered". . . . . He offered Himself
We surrender to death . . . . He embraced it
We relinquish the spirit. . . . . .He dismissed it
We may choose the time to die. . . . . He chose to die
We can only shorten our lives. . . He merely suspended his life, only to re-engage it at will
A few have raised the dead . . . . He raised *Himself*
Our death is passive, something we suffer . . . . . . His death was active, * something He commanded

He died *on* the cross but not *from* it.
He may even have died
with a ruptured heart,
but not because of it.

All these things were both possible and meaningful because, in Augustine's words, while it was "not impossible for Him to die," it was "possible for Him not to die". And these conditions of his life in no way disqualified his nature as truly Man, because these same conditions of life applied to Adam before he fell. The significance of the miraculous conception is that by this means Jesus Christ escaped the physiological consequences of Adam's disobedience, namely, the inheritance of physical mortality.

Unless this had happened, the Lord's death could only have been premature and not in any way vicarious. He would have been, as some have even suggested, merely the first Christian martyr. But his potential physical immortality certified that his death *was* vicarious. And yet his body was still truly a human body and He truly a representative of Man as the Last Adam. There is a circle here of cause and effect which cannot be broken – and the organic unity of the Christian Faith is critically involved in it.

Furthermore, He must not merely be Man, He must

* In the light of 1 Peter 3:18, it might reasonably be argued that He was "put to death in the flesh", making his death as passive as our death always is. If so, we indeed have a contradiction of all the evidence to the contrary. However, it is important to note that the Greek word *thanatoō* here rendered "put to death", can mean and is frequently translated "*condemned*, or *delivered up*, to die". It is so viewed in Mark 14:55 and Romans 8:36.

be God-made-Man. For whereas one man may die for another man on the principle of balanced compensation, were He only man – no matter how perfect – He could not have substituted for more than one man. Only by being more than man, and yet man nonetheless, could He make in Himself a *sufficient* sacrifice not only for my sins but for the sins of any man who will avail himself of it.

Moreover, the first man must not only have enjoyed the potential of physical immortality but he must also have possessed "original righteousness", by which is meant true moral freedom. The First Adam *need* not have sinned even as he *need* not have died, and thus the Last Adam was truly Man even though He never sinned.

Besides these things, Adam must have had a sense of moral accountability which made him a unique creature with a conscience towards God and the full capability of recognizing the nature of sin and the rationale of judgment. On these foundations was built a species, every member of which is capable of redemption and able to perceive the rationale of salvation as it applied to himself. Man is such a creature that he can by grace recognize his need of salvation when fallen and can embrace it by exercising the necessary saving faith.

These theological aspects of the biblical record of what happened to the two Adams, both as to their origins and their deaths, cannot be rationally integrated into an evolutionary world view *applied to man*.

Whereas it is true that the *application* of the redemptive process to the individual depends on the nature of man's spirit, it is also true that the *manner* of man's redemption has hinged upon the nature of his body. For this body was originally such as to permit the Son of God to be made Man in order to redeem man by his substitutionary death while at the same time in no way violating or surrendering his own divine nature. In short, man's sense of need originates in the unique properties of his *spirit*, and his redeemability hinges upon the unique properties of his *body*.

The theory of evolution applied to man makes a shambles of this Plan of Redemption. As Kirtley F. Mather observed in an article contributed to a volume of papers entitled *Science Ponders Religion* which were edited by none other than Harlow Shapley, "When a theologian accepts evolution as the process used by the Creator, he must be willing to go all the way with it". And I venture to say that no one can accept the evolution of man and still hold firmly in a truly rational way to the Faith once for all delivered to the saints.

In the very essence of its internal structure, the theology of redemption is challenged by evolutionary presuppositions and any satisfactory "wedding" of the two is logically impossible. The rationale of the Plan of Salvation is based entirely on the concept of balanced restitution, eye for eye, tooth for tooth, human life for human life. This simple fact lies at the very root of biblical theology.

H. G. Wells was perfectly correct when he wrote in 1920: "If all the animals and man had been evolved in an ascendant manner, then there have been no first parents, no Eden and no Fall. And if there has been no Fall, then the entire historical fabric of Christianity, the story of the first sin and the reason for the Atonement upon which the current teaching bases Christian emotion and morality, collapses like a house of cards". Such was the persuasion of a man who had no Christian convictions, but was more perceptive than many who have.

In a similar vein James Orr observed, "I do not think it can be sufficiently emphasized that Christian truth forms an organic whole – has a unity and coherence which cannot be arbitrarily disturbed in any of its parts without the whole undergoing injury. Conversely, the proof that any doctrine fits in essentially to that organism and is an integral part of it, is one of the strongest evidences we can have of its correctness". As will be seen in Chapter 18, it is a great pity that a number of other stalwarts of the Faith have not applied this test in their own thinking about the matter.

For many devout Christians today who have adopted evolution in place of creation, the problem lies in their unwillingness to extend the consequences of their broadened faith. They can only live with the substitution of evolution for creation because, while their knowledge of biology is often profound in many respects, their understanding of the organic unity of the Christian Faith has not been adequately worked out and they are unaware of the real nature of its logical structure and how impossible reconciliation really is.

POSTSCRIPT:

A single Old Testament passage sometimes foreshadows a whole series of events in the New Testament in a truly remarkable way. Thus in Exodus 12:5-7 we have the following words: "Your lamb shall be without blemish . . . . and the whole assembly of the congregation of Israel shall kill it in the evening . . . . and when I see the blood I will pass over you".

Accordingly, we have the Lamb identified as "the Lamb of God" in John 1:29, brought to the bar in John 18:30 and declared *de facto* "without fault" in John 19:4 and 12, having been brought by the whole assembly (Acts 4:27), and then "slain and crucified" (Acts 5:30) "in the evening": and finally, the blood sprinkled (Heb.9:11-15) that the judgment of God may pass over us.

Chapter 16

THE MAN WHO RAISED *HIMSELF* FROM THE DEAD

"In another form . . . ." *Mark 16:12*

It is a remarkable fact that the Early Church was far more concerned to celebrate the resurrection than the crucifixion. Illustrations of the resurrection were painted on the walls of the miles of catacombs under the streets and houses of Rome, from the very beginning. By contrast, the sign of the Cross is almost if not quite absent from them; the earliest known case is believed to have been done at a much later date by a pilgrim visitor.

Kenneth Clark observed that the Cross is hardly ever represented in earlier art forms, and when it is, it tends to be tucked away in the corner of the work rather than central to it. It was not until the fifth century that Christ was actually portrayed on the cross between two thieves.

It would, therefore, seem that the prime emphasis in preaching was on the resurrection rather than on the crucifixion, a fact which is a little surprising in view of the modern emphasis upon the Lord's death and the virtual neglect of his resurrection. William Barclay, whose commentaries are extremely popular but whose attitude towards the Scriptures is far from conservative, stated publicly that he does not see the virgin birth as having any essential place in the Christian Faith and is "mystified" by the resurrection.

In this chapter, I obviously cannot treat the resurrection in any sense comprehensively. My object is limited to drawing attention to a number of seemingly incidental remarks made by the four Evangelists which have a very specific bearing on the basic theme of this volume. What I wish to underscore is that the biblical record concerning the nature of Adam must be taken to mean what it has traditionally been taken to mean or the strictly logical coherence of the whole Plan of Salvation, *of which the BODILY resurrection of the Lord forms the capstone*, breaks down completely.

The amount of incidental detail which the four Gospels provide on the subject of the Lord's resurrection body is remarkable. Since *our* resurrection bodies are to be fashioned after the pattern of his and are to be an essential part of our personal identity in heaven, it is strange that one hears so little about the matter from the pulpit. After all, we shall spend eternity in them!

I want to discuss in this chapter what I believe to be the significance of some of these details in relation to the Lord's body under what appear to be three conditions of operation: (1) as it was when laid in the tomb; (2) as it was immediately after He raised it very early on that resurrection morning; and (3) as it was shortly afterwards, transformed into a new vehicle for the manifestation of the Lord's person with some entirely new and astonishing properties, properties which our resurrected bodies are also to share.

It will help to set the stage by listing briefly the basic facts which the New Testament provides in considering the circumstances surrounding this progressive change between the resurrection and the Ascension forty days later.

(1) We are told that *He* raised Himself: a circumstance unique in history.

(2) But we are also told that an angel, not the Lord Himself, rolled away the stone that sealed the tomb: suggesting, at first, real limitations to what He could do in that body.

(3) The body in which He first appeared to Mary Magdalene was evidently the very same body that had been laid in the tomb three days before.

(4) Contrary to what would be expected in the circumstances, that body had experienced no corruption whatever during its three days of entombment.

(5) A transformation of some sort then took place in that body between the first appearance to Mary and the next appearance to Mary and others, that seems to have occurred quite soon thereafter. This transformation endowed it with entirely new properties which, however, in no way left its identity in any doubt. It was still *his* body.

(6) What had transpired in the interval between those two appearances can be tentatively conjectured from certain passages of Scripture which, if the proposed interpretation is allowed, shed light on the circumstances surrounding (2) and (3) above.

(7) The significance of the very deliberate way in which the Lord at first protected Himself against being touched but afterwards invited, indeed commanded it, is examined in the light of certain statements in Hebrews not usually related to this seeming contradiction.

(8) The importance of these extraordinary new properties of a resurrected human body, as they relate to our own future, is then explored. They suggest that our bodies, and therefore *we*, will enjoy a new kind of freedom from present limitations while yet retaining our identity fully and unequivocally throughout eternity. The prospect is a wonderful one, and the importance of the human *body* is magnified tremendously. Its potential as a vehicle for the redeemed spirit of man to express itself effectively and actively in the New Heaven and New Earth that is promised, is seen to be well-nigh inexhaustible.

I am well aware that the Gospel records of the resurrection appear to many to be so full of contradictions as to be almost beyond reconciliation. It would be a foolish man who imagined that he has the final key after the attempts of all others have brought no universally accepted resolution. All I hope to do in this chapter is to show that we are provided with a chain of circumstantial details in these records which seem to me to move forward progressively to a single objective. This progression demonstrates a significant change which took place in the nature of the Lord's body at some point between the time of his first appearing to Mary Magdalene and his ascension from the Mount of Olives, and a reason why the change was not immediate is suggested.

This "progressive revelation" as it were, is not materially affected one way or the other by doubts as to who saw what, or even in what precise order *some* of the incidents took place. I venture to say that the following very summary statement will not be disputed by those who take the Word of God to be entirely free of error. One is led to conclude that even the very wording used has been inspired. It seems to me that a great deal hinges upon the assumption that the wording is indeed to be taken very seriously. In short, these are not merely intriguing incidents. They are designed to communicate a truth about the nature of the Lord's resurrection body which is of profound importance, and perhaps could not have been made clear in any other way. Here we have the pattern after which *our* bodies are also to be fashioned, though not progressively as His was.

I propose to put this summary statement in the form of a series of questions, each of which refers to a particular circumstance that may not have struck the reader previously as of much importance. I have to assume, in the interests of brevity, that the reader is more or less familiar with the broad sweep of events and therefore with the general background of the particular details which constitute the focal point of each question. Here are my questions:

(a) Why was it necessary for an angel to roll away the stone (Matt.28:2)? Why did the Lord not simply pass through it, or even roll it away Himself by his own mighty resurrection power?

(b) When He appeared to Mary Magdalene, the first to come to the tomb, why did He *forbid* her to touch Him (John 20:17)? What did He signify by his explanation of why such contact was not permitted?

(c) When Mary and her companions arrived back at the tomb later, why did He now *allow* her and her companions to hold Him by the feet (Matt.28:9)? For this was certainly making physical *contact* in a very concrete way.

(d) Why did the two who walked to Emmaus not recognize Him visually as Mary and her companions had recongized Him, until He performed the act of breaking bread (Luke 24:30)?

(e) Why are we told that the doors were locked, which seems too obvious a thing to need mentioning since it was night-time?* Why is such emphasis placed on his means of self-identification? And what did He mean precisely by the phrase "a spirit hath not flesh and *bones* as ye see me have" (Luke 24:38,39)?

(f) Was there a specific reason in the purposes of God for the absence of Thomas at his previous appearance and his presence now eight days later (John 20:25,26)?

(g) Is there any particular reason why the climax of these personal appearances (John 21:1-14) should once again have involved a meal of which the Lord was this time not only the *host* – but a host whose invited guests were asked to supply part of the meal themselves?

It will be seen from what follows that this chain of events falls into a pattern, each link making its own particular contribution, a contribution usually crystallized by a small circumstantial detail the significance of which is easily overlooked. I am not making any attempt to address the many apparent contradictions that have been remarked upon by those who have made a far more intensive study of these events. But I do not think the resume which follows will be seriously challenged as to their sequence or the chief

---

* The word translated "shut" implies locking in the Greek and forms the root of their word for "key".

characters involved in each instance. Having said that, let me boldly plunge where even angels might fear to tread.

The body of any condemned man was removed before sundown to be buried according to Jewish law (Deut.21:23). But no corpse was given *full* preparation at once but merely protected against attack by insects and wrapped for the time being. The tomb was then temporarily closed to protect against predators. It would be re-opened after three days.

The reason for not completing the burial arrangements at this time was that "certification of death" was not granted until the "third day" because by this time it was assumed that a certain marked change would have taken place in the appearance of the face which, by Jewish tradition, signified that the spirit had left the body permanently and would not return. The object of this precaution was to ensure that death had really occurred and that no natural recovery would take place as sometimes happens even in modern times, even after certification of death.

Those who performed these burial rites were usually women rather than men, and were customarily personal friends and relatives of the deceased. This circumstance accounts for the fact that a succession of women came to Jesus' tomb very early on the third day with various spices to care for the body for the last time before final closure of the tomb.

When Mary Magdalene, who was the very first to come, arrived while it was still dark, she had every expectation that the tomb would be *closed* but evidently wanted to be there as soon as she could for the opening of the tomb and certification of death according to law. It was the last sight of their beloved Lord any one of them could expect to have as the body was prepared for final interment.

Remember that it was still dark (John 20:1), but not too dark for her to find something that surprised her greatly. *The stone had already been rolled away.* The woman cautiously approached the open tomb, it was all so quiet and so dark and there was no one around to give her assurance, she did not dare to go in but went back to find more company.

She returned, now accompanied by "the other Mary" and Salome, the mother of James. By this time it was just beginning to get a little lighter (Matt.28:1). To their amazement the angel who had rolled away the stone (Matt.28:5,6) was at that very moment sitting upon it!

The angel spoke to them quite naturally and invited them to examine the tomb and see for themselves that the Lord was no longer there. It was indeed empty, but inside were two other angels who actually rebuked them for their incredulity!

"He is not here, but is risen", they said. "Remember how He spake unto you when He was yet in Galilee, saying, 'The Son of man must be delivered into the hands of sinful men, and be crucified, and the third day rise again'" (Luke 24:6, 7).

We do not know where Mary Magdalene's companions went – perhaps to take home the spices they had brought, not knowing what else to do! But we do know that Mary ran at once to tell Peter, and John the disciple whom Jesus especially loved (John 20:2-6). Peter and John (with Mary never able to keep up with them!) immediately ran to the tomb to see for themselves: and their hurried journey is described in one of the most marvellously descriptive passages ever penned in so few words!

"So they both ran together: and the other disciple did outrun Peter, and came first to the sepulchre. And he stooping down, and looking in, saw the linen clothes lying; yet went he not in. Then cometh Simon Peter following him, and went into the sepulchre, and seeth the linen clothes lie, and the napkin that was about his head, not lying with the linen clothes, but wrapped together in a place by itself. Then went in also that other disciple, which came first to the sepulchre, and he saw, and believed" (John 20:4-8).

In the tomb they saw the linen clothes lying in one place and the head wrapping carefully laid by itself, evidently folded and laid down as though the Lord's body had been gently unwrapped. Certainly there was no evidence that the body had been stolen, for such care would never have been taken by thieves!

Meanwhile, Mary followed breathlessly some distance behind and only arrived as Peter and John were about to go back and tell the rest of the disciples what had transpired. The body was gone, and so had the guards, but it must have seemed obvious to them that it was not the work of thieves for this false report was not circulated by the disciples but by the Lord's enemies (Matt.28:12-15).

Meanwhile, Mary remained at the entrance to the tomb, distraught by what they had discovered, and weeping at her personal loss. She could not even perform the last rites to the body of the Lord whom she had loved so dearly and understood so well (John 12:3).

Looking into the tomb through her tears, she suddenly saw two others were there, one at each end of the place where the body had lain. They said to her,

"Why are you weeping?"

"Because," she said, "they have taken away my Lord and I know not where they have laid Him".

And having said this, she turned away from them and saw someone else in the opening, his sihouette sharply outlined against the dawning sky as she stood in the tomb. This one said to her (John 20:15),

"Why do you weep? Whom do ye seek?"

This, she thought, must be the gardener – for after all Nicodemus, who had just purchased the property and had had the rock tomb excavated, was a rich man and very probably had a gardener to look after it. So she said to him,

"Sir, if you have borne him hence, tell me where you have laid Him and I will take Him away".

Seen against the lightening sky and through her tearful

eyes, she did not recognize who He was until Jesus did the one thing guaranteed to identify Him to her unequivocally: He simply spoke her name.

"Mary!"

Mary fell instantly on her knees before Him and would at once have held Him by those beloved feet had He not said quickly,

"No, Mary, you must not touch Me!"

And then He told her why:

"Because I have not yet ascended to my Father. But go to my brethren, and say unto them, 'I ascend to my Father and your Father; and to my God and your God'" (John 20:16-27).

And I can imagine her, almost beside herself with joy and, asking no more questions, running breathlessly back to where she knew the disciples were gathered, to tell them that she had actually seen the Lord! He was alive! He was risen indeed! She had spoken to Him face-to-face!

In the meantime, while these events were transpiring and the message was being eagerly shared and talked about, two of the Lord's friends, Cleopas and his wife, were walking soberly back home to Emmaus (Luke 24:18). For although the news was spreading quickly no doubt, almost no one really believed (save for Mary Magdalene) that the Lord was indeed risen in body and was back among them once more . . . .

This "walk to Emmaus" is perhaps one of the most dramatic scenes in the whole of Scripture! The two travellers as they go talk about all that has happened in the last few days: the dashing of their hopes that Jesus was indeed the promised Messiah, and how He had been so cruelly treated, and unresisting had been led away to die the most frightful of all deaths on a cross. And now there were these rumours . . . .

And as they walked, Jesus came up behind them and walked a little apart but still close enough to hear their conversation. He soon joined them unobtrusively and asked them what they were talking about so earnestly. They were naturally surprised at his ignorance, but He in turn rebuked them and asked why they hadn't realized that it had all been foretold beforehand!

"O fools, and slow of heart to believe", He said. And so beginning at Moses and all the prophets, He expounded unto them in all the Scriptures the things concerning *Himself* (Luke 24:18-27), thereby explaining the meaning of what had been happening.

In no time at all it seemed, they were home, and they turned aside to go into their own house. Their newly-found companion, still unrecognized, went on as though He would not invite Himself in. But they urged Him to come in and He accepted the invitation. Together they went in and very soon were sitting at table having refreshment. Meanwhile, the conversation continued, until He took a piece of bread, blessed it and break it – and suddenly their eyes were opened and they recognized Him. And no sooner did they recognize who He was than He vanished out of their sight!

Now *why* had they not recognized Him before? Perhaps the answer lies in Mark 16:12. "He appeared to them *in another form*" (Greek: *en hetero morphe*).

If we were to find another word for this phrase, we might say his body had been transformed. In fact, He had now assumed a transfigured form – it was his own body still, but somehow wonderfully changed. The facial agony was gone, the shadow of the three hours of darkness was dispersed, the marks of the crown of thorns and the buffeting, the bloody sweat, the gasping for air and the desperate need for water in the mouth, the matted hair and the bowed shoulders that could not sustain the load of the cross – in short, the "tragic" figure of an utterly abused and condemned victim in whose face there was "no beauty that we should desire Him" (Isa.53:2) – all that was gone. *That* had been their last impression and their only remembered one. But this stranger who shared their walk back to Emmaus was utterly unlike the Lord as they had seen Him barely three days ago.

And it is almost certain they had no expectation of ever seeing Him again, least of all vibrantly alive and so beautiful. It is really no wonder they did not recognize Him, for He did indeed appear in "another form", a figure of beauty not shame. In fact, so different was this body that He could share a meal with them, eat their food, *and then suddenly disappear, vanishing without a trace – not even a trace of the food He had eaten*!

No wonder they arose at once and went back to the city to join the other disciples. The Lord was alive indeed! The Lord had risen, as He had promised!

Now, the question arises: If He could vanish so easily now, why was it necessary for an angel to come and roll away the stone so that He could leave the tomb – as the stone had been rolled away for Lazarus to come forth? Why would He not simply pass *through* it?

Is it possible that the metamorphosis of his body did not actually occur until *after* Mary Magdalene had seen Him? If so, must it not be that the body which was laid in the tomb, his physical, earthly, and terribly wounded body, was the very body which He said He would raise up – that very temple (John 2:19), and nothing less?

But such a body could not pass through the stone closure! Nor would it be *humanly* possible, even if it were resurrected and given new life, for such a body to roll the stone away, blocked by a wedge in the track as it certainly was, on Pilate's orders. From inside, no one, not even a Samson perhaps, could have forced it over the impediment. Someone from outside had to remove the "lock" and roll it away, whether man or angel.

Thus we might, not without reason, reconstruct the scenario and suppose that the Lord, the Jehovah of the Old Testament, did indeed revitalize his own Temple, but not yet to change it in any significant way. An angel rolled away the stone, and two other angels removed the wrappings —just as all this had been done for Lazarus. Only then did He step forth, his "old" self, still bearing the marks of the wounds indeed, and still clothed in his earthly body. As Mary Magdalene returned, perhaps He stood aside in the dim light until she had entered the tomb, and then stood silhouetted against the sky as she came tearfully and wonderingly out. The rest we know.

But a question remains. When *did* the transformation occur? Why did He not permit her even to touch Him? Was there a task He must yet complete while still clothed in this *earthly* body? I believe there was.

He had performed on earth the first part of his sacrifice "in his own body on the tree". As both High Priest and sacrificial Lamb He had yet one more duty to perform "in his own body", this time in heaven. He was now about to ascend to His Father and his God to complete the sacrifice by personally presenting his own blood.* *This* ascension was quite different from the ascension at the end of the forty days. This was an ascension to be followed by an immediate return to earth. Did not the angel say to Mary and to her friends a little later, "Go your way and tell the disciples and Peter that He goeth before you into Galilee; there shall ye see Him as He said unto you" (Mark 16:7). Is there any evidence from Scripture that the Lord really did ascend for this purpose to his Father and his God? I think there is.

Hebrews 9:11 and 12 reads: "But Christ being come an High Priest of good things to come, by a greater and more

---

* **Because his body, unlike our bodies, had not seen corruption in the grave, we can be assured that his was the blood of a Lamb without blemish and without spot. He was, as we have seen, made only in the *likeness* of sinful flesh, not sinful flesh in fact (Rom.8:3). Thus his blood was likewise uncorrupted and could avail for an atonement. This is why Peter said so pointedly: "*He whom God raised again, saw no corruption*; be it known unto you *therefore* that *through this Man* is preached unto you the forgiveness of sins: and *by Him all that believe are justified* from all things, from which you could not be justified by the law of Moses" (Acts 13:37-39). In short, our very salvation hinges upon the fact of that un-blemished body and uncorrupted blood.**

perfect tabernacle not made with hands, that is to say, not of this building [i.e., his body: cf. John 2:19]; neither by the blood of goats and calves, *but by his own blood He entered in once into the Holy Place*, having obtained eternal redemption for us". Here, I suggest, He laid before heaven, before *his God* and his Father, before the angel hosts, and even before Satan perhaps, the blood of the everlasting covenant that has obtained for us eternal redemption.

I do not suggest that there is any way of explaining such things in physical terms, but it seems to me that the simplest explanation of why the same Mary Magdalene (and her friends) was allowed to touch Him and hold Him by the feet upon the *next* meeting with Him, is that in the interval He had safely fulfilled the function of High Priest and completed the sacrifice of Himself as the Lamb of God. He, the great High Priest had offered Himself, the Lamb. Under the old Covenant any human contact with the High Priest instantly defiled the priest and disqualified him, so that a "reserve" priest was always kept in readiness on the Day of Atonement in case the appointee for that year should accidently be defiled by any such contact. Mary could not touch Him until He had completed this part of his sacrifice. Thereafter He not only allowed contact with his body, but invited it, and on one occasion even commanded it!

Yet we know from the next appearance that while the transformation had restored his beautiful face (cf. 2 Cor.4:6), the identifying marks of the wounds of crucifixion in his hands and feet and of the spear in his side still remained. Perhaps this will serve throughout eternity to show that whereas our bodies must be dissolved and replaced, *such total replacement was not required in his case*. For his body had remained without internal blemish of sin. Presumably, the Emmaus couple (Cleopas and his wife) had not noticed his hands. But his body had been changed in one very significant way: it was his body still and identifiably so, but it was no longer his earthly body as to its functioning principal. Nevertheless it was still *His* own body.

Which brings us to his first appearance among his disciples gathered together with all the doors shut, for fear of the Jews (John 20:19). Unlike the closure of the tomb which could only be opened from the outside, these doors were "locked" from the *inside*. But Jesus passed through them and suddenly appeared in the room where they were all gathered discussing what a few of them had just seen and heard. When He suddenly became visible, the very circumstances of his appearing must have convinced them all that it was a mere ghost of Himself. To prevent this misapprehension, and without hesitation, He held out his hands and showed them his feet and said, "It is I myself, handle me and see; for a spirit hath not flesh and bones as ye see me have" (Luke 24:39).

The visual impression of this display of the marks of the crucifixion must have been tremendous, even though it was as yet only visual and not tangible evidence. Even a ghost might *appear* to have flesh and bone, for how else could a ghost's hands appear as hands? But clearly, being hesitant to put Him to the test by actual contact, the disciples still stood in some doubt.

Almost in rebuke the Lord said, "Have you any food?" And someone gave Him a piece of broiled fish and a piece of a honeycomb (Luke 24:39-43). We are then told, simply, that "He took it and did eat before them".

This was a beautiful example of the Lord's dealing with any complex situation. The alternative was to *persuade* each one present (perhaps up to 20 persons) to personally examine his hands and feet and side. How otherwise could all present be individually convinced? But by this simple means He left all who had eyes to see in no doubt whatever as to the reality of his bodily presence.

They must have waited until He had actually swallowed it, and then perhaps *He* waited silently for them to recover from their amazement! Surely there could be no longer any doubt that it was really Himself and that *his body was as real as the food which He had eaten*. And yet He had come through closed doors and appeared out of nothing in their midst! This was indeed a new kind of embodiment . . . . manifestly substantive, and yet not subject to any physical barrier.

But it should be noticed that He had said, "A spirit hath not flesh and bones as ye see me have": flesh and *bones*, not flesh and *blood*. The Lord thus employed language which is not found anywhere else in the New Testament, although the phrase "flesh and *blood*" certainly is. It is used, for example, in Matt.16:17; 1 Cor.15:50; Eph.6:12, Gal.1:16; and Heb.2:14; and in every case it is obvious that the meaning is precisely to signify an *earthly* body. "Flesh and bones" would seem to signify a different kind of body, a body indeed, but a body differently constituted.

The Lord had avoided the word *blood*. Could it be because his body was now sustained by a different form of energy, vitalized by a different principle of life? The principle of his old life sacrificed on Calvary, namely, the blood, was now in heaven; and the new life was independent of that blood. In other words, the life which is in the *blood* had really been given and was never to be taken back again. It was in the realest sense an eternal sacrifice. Nevertheless, in order that He might henceforth remain Man, a human body was still necessary, a real body of flesh and bone, a functioning body; and yet a body now operating on some entirely different principle.

Eight days later He again appeared in their midst passing freely without any hindrance through closed doors as before. But this time Thomas was present. With his usual greeting He said, "Peace be unto you". And then turning immediately to Thomas and without asking him what kind of proof would satisfy his lack of faith, He commanded him to do exactly what Thomas had said to the rest of the disciples would be the only proof he could accept (John 20:24 and 25), namely, to assure himself that the nail prints were *real*. It is obvious that the Lord had been present, though invisible, when Thomas had made this statement, whereby

He showed them that his presence was just as real when invisible as it was when visible – and one has to assume just as *bodily* real, even though they did not know He was there. The reality of his new body did not depend upon its visibility.

It would appear from John 20:27 ff. that Thomas was astounded to hear the Lord command him to satisfy himself on his own terms. We are not told that he responded by actually doing it – only that he said, perhaps somewhat breathlessly, "My Lord and my God!"

It is clear therefore from these records that the Lord would have submitted to any test that any solid live human being might have asked of Him to prove the reality of his body, simply because his body was capable of satisfying any such test. It was a real body, more *real* than ours in fact, because now indestructible.

The final episode is truly astounding. It took place probably not long before his ascension. It is recorded in John 21:1-14. The circumstances are as follows.

The disciples had fished all night without success. As the day dawned and they were near to the shore, they discerned someone standing at the water's edge. Apparently there was more than one boat involved in the fishing party. Across the intervening water a voice that they may not at first have recognized called out to them,

"Children, have you any meat?"

"No", they answered Him.

"Cast your net on the right side of the ship", the stranger called back, "and you shall find".

Although it seems unlikely from what followed that at that moment they knew it was the Lord, for some reason they obeyed unhesitatingly. John writes: "And now they were not able to draw it in for the multitude of fishes".

John, the perceptive one, said to Peter in a low voice, "It is the Lord!" And as soon as Peter recognized the truth of it, he put on his fisherman's coat in order to cover his nakedness, jumped into the shallow water and ran towards the shore. Meanwhile, the other disciples in another little boat drew near; and between them they dragged the net, so full of fishes, to shore. And then they saw a fire of coals there, and fish laid thereon and bread (John 21:9).

It is hard to imagine what this really signified. Had the Lord *caught* these fish? Had He *baked* this bread? Had He *gathered* the wood to burn, and *lit* the fire to form the bed of coals? Did He go into the water and catch the fish, and search along the shore for the wood? Or did He create it all, fish and bread and coals? Somehow, one likes to think He had gathered them. Despite the transformation that had taken place in his body, He was no less a man – and now He was also a host.

The surprising thing is that He apparently, and I am sure quite deliberately, did not actually gather enough fish for the number of guests who had arrived! And so He said, "Bring some of the fish which ye have now caught" (verse 10). Surely this was *not of necessity* but was rather a beautiful gesture, guaranteed as nothing else would have done, to give them a sense of reality in what was happening, by having them contribute something more than merely their presence. For everyone of them must have felt almost paralyzed in amazement by what was taking place. Yet Peter had so recovered himself that he went over and pulled the net in a little further still and took time to count the great haul of fishes they had just made; 153 of them! He even noticed that the net had not broken and would need no repairs . . . .

By this time the meal was cooked and Jesus said, "Come and dine" (verse 12). It is an unimaginable scene, or perhaps one should say rather an uninventable scene, and everything that was done and said was entirely appropriate to the setting. Silently they all sat down around "the table", and we are told that none of the disciples dared to ask, "Who art thou?" They *knew* who He was.

And one wonders how the conversation went as they

shared this picnic by the sea, to which in his graciousness they had been allowed to contribute their share. And Jesus broke the bread and gave it to them, and the fish . . . .

And how many thoughts must have gone through their minds! Perhaps they recalled the feeding of the 5000 . . . . and the 3000 . . . . and the last supper. That supper was no longer the last one. *This* really was the last supper, a supper of joy, and peace without shadow.

Quite properly, the curtain is drawn on that scene by saying, "So when they had dined" (verse 15). And what a dining that had been!

The reality and the potential of that wonderful body had been amply demonstrated. It had served Him superbly. It had graced his earthly life. It had never seen corruption even in the tomb. He had dis-engaged Himself from it for a few hours and re-engaged it to complete his sacrificial work. He had borne our sins in it, on the cross. He had glorified the Father in it. Men had seen God objectified in a body which under no circumstances had ever displayed any cause for shame.

Now, it had received the final and higher form for which it was created. It was set free at last from all vulnerabilities and all limitations. It had achieved immortality in the most absolute sense. It would never be laid aside again and therefore the Lord Jesus would remain for ever truly God incarnate as Man. In it He could do whatsoever He wished to do, and there were no barriers to its movement or functioning, and no waning of the energies which empowered it.

Such, then, was the potential of the body of Jesus Christ, the Last Adam; and such, then, must have been the body of the First Adam until he almost but not quite destroyed that potential for ever, by eating the forbidden fruit and poisoning himself and the human race into a mortal state.

Is it conceivable that such a glorious vehicle for a truly

human spirit could ever have had an animal origin? Surely only a miraculous conception can account for the body of the Last Adam; only direct creation can account for the body of the First.

Evolution cannot allow that an animal's *body* has any transcendental significance, and since it insists our bodies are essentially no different, it cannot allow that *they* have any transcendental significance either. As far as our bodies are concerned, they die like the beast's. Only a ghost of ourselves can be allowed continuance and most evolutionists would flatly deny even this.

But the Scriptures speak otherwise. We are assured of re-embodiment, in a body like his glorious body, sharing its incredible potential, and no longer being a barrier to the spirit but a perfect vehicle for its expression.

I wonder if the Christian evolutionist has *really* worked out his own faith in the light of the Lord's bodily resurrection . . . .

Chapter 17

WE, TOO!

Speculative? Of course it is! As one whose professional life was spent as Head of a research laboratory in Human Physiology, I know that one can make little or no progress without it — only we call it hypothesizing. And it has proved highly successful in the laboratory.

Why should not the Christian exercise freedom of mind in the same way — only this time with a mind renewed and a mind informed and channelled by revelation and disciplined in the use of logic? Why not!

A human being is *not* a human being without a human body. Unless the body is constitutionally part and parcel of man's being, the emphasis in the New Testament upon the resurrection of the body is quite unaccountable.

If we can be *persons* without a body, there seems to be no reason why Paul should have argued that unless the body is raised we are of all men most miserable: nor why he should have said our final expectation is 'the redemption of the *body*'. The proof of the Lord's resurrection was not his re-appearance as a ghost but as a person with a body of flesh and bones. Resurrection is not merely spiritual survival but bodily survival as well.

I don't, however, wish this chapter to be an anticlimax in the form of a catalogue of passages of Scripture proving the point. This has been done many times by detailed exegesis of the wonderful assurances of the reality of the next world – such as we find in 1 Corinthians 15:35-38; Philippians 3:20, 21; and the final chapters of Revelation. I'm not attempting to prove the resurrection of the body: I'm starting with it as a basic premise.

What I want to do is to exercise some imaginative freedom and reflect upon four aspects of the potentials of such a bodily existence as the Lord Jesus Christ experienced and still experiences as a Man in heaven, for here is our pattern according to the promise of the Scriptures.

These four aspects may be summarized as follows:

(a) *What shall we* **be**?
(b) *What shall we* **do**?
(c) *What shall we* **know**?
(d) *How shall we* **be recognized**?

What will it be like to be entirely free for ever – free to be what we would like to be, to do what we would like to do, to know what we would like to know, to go where we would like to go, and to meet whoever we would like to meet.

How will we feel when we suddenly realize we are incorruptible and immortal at last, without fear or pride? What will it be like when we are free of gravity and of the limitations of space and time: free of need, free of hunger, pain, delay or impatience, hatred or malice, boredom or unfulfilled aspiration or unwanted partings – of all and any of the sorrows and disappointments that characterize this life and turn a beautiful earth into such a blessed vale of tears.

Name *anything*, absolutely anything that is a source of human anxiety or grief or shame, and *that* will be absent from our world. What freedom this will bring, and with what safety it will be enjoyed! The real reason we cannot be allowed even the good things we long for, including fruitfulness in his service, is that all too frequently it would not be safe for us. We would all too soon be plagued with spiritual pride if our labours were to be as blessed as we would wish. Such is the paradox of Christian life because of that most troublesome of all sins – spiritual pride.

Let us, then, take a look at these four aspects of existence which will almost certainly apply to a state of freedom that is nevertheless *not* a state of disembodiment but of being embodied "gloriously" as promised in Philippians 3:21.

(a) WHAT SHALL WE BE?

We have the Lord's promise that when He returns He will not only receive us unto Himself (John 14:3) but that when we see Him *we shall be like Him* (1 John 3:1). This is what we are to BE.

The change will be instantaneous, "in a moment, in the twinkling of an eye" (1 Cor.15:51, 52). We shall be raised incorruptible and immortal for that meeting (1 Cor.15:52, 53). This diseased and decaying body will be transformed "that it may be fashioned like unto *his* glorious body, according to the working whereby He is able even to subdue all things unto Himself" (Phil.3:21).

And if you have questions about the 'how', I suggest you read 1 Corinthians 15:35-50. The words scarcely need comment: they are clear and unequivocal. What will rise from the dust is to be an outgrowth of what has been sown. To quote Thomas Boston again, "There is a vileness in the body which, as to the saints, will never be removed, until it be melted down in the grave, and cast into a new form at the resurrection to come forth a spiritual body".

Nor will there be any loss of identity, a fact which gave Job great assurance and led him to exclaim: "Though after worms have destroyed my skin, they will then also destroy this body, yet in my flesh shall I see God: whom I shall see for myself and *my own eyes* shall behold, and not someone else's" (Job 19:26, 27). We shall be ourselves and all experience will be firsthand.

God will give to each of us a body as it shall please Him (1 Cor.15:38), *having wrought us in the first place "for this very thing*" (2 Cor.5:5). It is far better that He should choose for us the particular form of embodiment that will make us whole again. As we shall see in dealing with (d), we shall really not be in a position to choose for ourselves. What He chooses for us will be a body that is perfectly appropriate to house the spirit which He has perfected for it in order to re-constitute us as the kind of person we have always longed to be both outwardly and inwardly.

Above all, *we shall be human beings, not angels*. The point is an important one. To Adam was committed the government and cultivation of the resources of the earth. He and his descendants were to multiply and fill it in order to "occupy" it in the proper sense. When He said, "Occupy till I come" (Luke 19:13), He implied two things: first, that He is coming again; and second, that He is coming again to "occupy" the earth as the Second Adam – with all that this implies.

If God committed to man the management of his created world, He had of necessity to provide him with the means. He had to provide him with a *mind* that would enable him to understand His will. And He had to provide him with a suitable *brain* in order that his *mind* could exercise its will upon the physical world. It would be foolish to suppose that, if there is to be a new heavens and a new earth, we would not continue to be provided with the same two pre-requisites.

We do not yet know how mind or will can act upon the body and use it as an extension of itself – as the hand, for example, puts our wishes into practice. Just how a spiritual force like my will can move a material object like my hand but cannot move the hands of the clock on the wall except indirectly, is a mystery. But certainly we have the will and the skill increasingly to manage the physical world if we were to set our hearts to do so.

But angels have no such direct powers to influence physical objects except by a temporary embodiment. The angel that "rolled away the stone" had to move a physical mass of perhaps 1500 to 2000 pounds. Even the little stone placed in the track and sealed in position by mud or clay so as to show whether it had been tampered with, would prevent its being rolled back unless it also were removed. No doubt the angel did this as a requisite first step. But it seems that the angel must have been temporarily embodied with the same kind of embodiment involved when the two angels took Lot and his wife "by the hand" and hurried them out of the doomed city of Sodom (Gen.19:16).

Yet angels are *not* men, because by definition a man is only man in a truly human body, and though angels may be temporarily embodied it is certainly not their customary constitution. Moreover, unlike man, they seem quite able to be fully conscious as pure spirits without bodies.

When the Lord Jesus became Man, He did so by becoming *flesh*, of human seed, of the seed of the woman and of the seed of David and of the seed of Abraham. He became partaker of flesh and blood (Heb.2:14), though without our defect. He did not take upon Himself the nature of angels (Heb.2:16), because He came specifically to act upon our world and to do it as we do it, not as an angel might do it. He lived among us, healing the sick, feeding the hungry, raising the dead, and sharing our physical life to the full. Moreover, He came to die, which God as pure spirit cannot do.

Thus man by his very nature can interact with the physical world. And because his spirit and mind can comprehend the will of God, he can fulfil the will of God within this physical framework as its governor. He is a link across two realities, the reality of the physical world and the reality of the spiritual world. This is what he is – a link and a medium; and in the new heavens and the new earth since our basic constitution of spirit and body is to be retained, *we shall continue to be a link and a medium.*

(b) WHAT SHALL WE DO?

We are not told very much about the nature of the new heaven and the new earth except that they shall *remain* and therefore will not wear out or run down. They are to be *new* and unshakeable (Heb.12:27) which may perhaps signify that they will be stable and not subject to catastrophic disturbances such as astronomy and geology hold the present universe to have been subject to in the past.

The present world seems clearly to have been designed as a habitation for man as he is now constituted. Indeed, this is not only true of the earth but probably even of the heavens also, i.e., of the whole universe. This has in recent years been the considered opinion of a number of prominent cosmologists.

It would seem reasonable to conclude that if we are to undergo a transformation of a certain kind and to be placed in a *new* universe, that universe will accordingly be a transformation of this present one. It will therefore be *new* in this sense: not merely a replacement in the same form but a re-formed replacement in which we shall live and move and have our being. In other words, we shall belong in the whole of it – as we cannot be in the present one, for lack of time if nothing else! There will then be the same kind of correspondence between our constitution and its constitution as there is at the present time, except that we shall not be bound by those physical limitations which currently bind us.

While this present heaven and earth will pass away (Matt.24:35), the new one will not (Isa.66:22), and neither shall we. Thus the words, "Of the *increase* of his kingdom there shall be no end" (Isa.9:7), may take on an entirely new meaning, the key word being *increase*. A few years ago such a prospect might have seemed utterly absurd, but not any more – only we shall not need space ships nor space suits, and our movements may even exceed the speed of light!

I do not for one moment anticipate that our "time" will be spent playing a harp, much as I would like to be able to play *any* instrument well! But I believe such an achievement would be easily within our reach just by willing to do it. So likewise, I'm confident we shall be able to roam the earth, or the sky, and indeed the whole universe, at will. Yet I suspect it will not be an idle roaming.

There will be creative responsibilities, 'rulerships' as it were, over whatever will correspond to the "many things" of Matthew 25:21. Whatever the reality may prove to be, I am sure it will be easily recognizable as a fulfillment of our capabilities when the time comes. And my prediction is that the sense of reward we get in *this* world for an achievement well done is a harbinger of a sense of far greater achievement for things yet to be accomplished in *that* world (John 14:12). Man was not designed for idleness.

One final point. Such 'doings' assume the continuance of our embodied humanity with its potential for creative activity. But this seems to assume also that the Lord will retain *his*: and this I believe He will. The mark of his humanity was (and is) his willing *subservience* to his Father (John 6:38), even as the mark of his deity is his rightly claimed *equality* with his Father (John 10:30). If this is a reasonable assumption, then perhaps we have assurance of the permanence of his humanity from the fact that when He shall have had all things (*tapanta* in the Greek, i.e., "the universe") subdued under his feet, He in turn will also Himself be subject unto Him (the Father) that put the universe under Him, that God may be all in all (1 Cor.15:28). He will therefore, it seems, never cease to retain his two natures, one of which we now also share. Neither He, nor we, will ever be in a disembodied state again, and this implies we shall never be without something to *do!* The fact that we are to be embodied can only mean that the universe will be 'substantial' and that we shall be able to act upon it.

(c) WHAT SHALL WE KNOW?

I suggest *we shall know anything we need or desire to know and our knowledge will be without error*. Since it seems impossible to contemplate the failure of memory, whatever we have once learned in that new universe we shall never forget.

But would not this imply an interim ignorance, and is not ignorance a kind of sin? From the New Testament, we know that the answer to this must be in the negative.

The Lord Jesus as *Man* did not know everything automatically and from the moment He was born. He did know everything He needed to know at each stage of his life, and all that He knew He knew perfectly. Of some things He seems clearly to have been ignorant. This was part of his self-abasement in assuming human nature, which in order to be truly human required that He lay aside or sublimate his divine omniscience sometimes. Thus while his occasional ignorance was real enough (as when He could not tell his disciples the time of his returning), it was not an ignorance due to sin.

On a number of occasions we see this laying aside of omniscience. For instance, we learn of his *surprise* at the fig tree which lacked the fruit He had expected to find on it, and of his great delight in the man who, though not a Jew, had a wonderful faith in his power to heal. On another occasion, He asked his beloved and bereaved friends, Martha and Mary, to show Him where Lazarus had been buried. Was this merely an accomodation to his friends in which He concealed his knowledge or was it a real case of limitation? It seems to me that He would not pretend under any circumstances, and therefore that He really did not know.

I believe in our perfected state in that new universe we shall still have much to learn. We shall know anything we need to know and learn everything we desire to learn. We shall never suffer from needless or undesired ignorance. Things we want to know, whether necessary or not, we shall know by some process perhaps akin to intuition, or merely by asking.

And as to asking questions . . . . With a direct line of communication involving neither delay nor hindrance of any kind, perhaps what *computers* are beginning to make possible for us in *time*, that which will constitute *brain* in our new bodies will do for us in *eternity* when channelled into the right data-bank. That heavenly resource could well be nothing less than the mind of God.

(d) HOW SHALL WE BE RECOGNIZED?

It would seem a simple matter to house the resurrected spirit in a recognizable body so that we could all know one another as we have known one another in this world. The size and shape and mannerisms of our bodies become part of our identity to those around us. And our facial features and expression clinch the matter. So we have plenty of clues by which to recognize one another.

But then a problem arises. What if it has been twenty-five years since we last saw some particular friend in this world? Faces change, and so do figures! How, then, shall we recognize each other if in the interval we have grown from infancy to maturity, or even from middle age to old age? In short, what stage of our life will our resurrection bodies reflect?

And even more problematically, let us suppose that a child is a year old when his mother dies. He grows up to be a well-known Christian leader and in due time, well on in years, he goes home to be with the Lord. How, now, will mother recognize son, or son mother? Adequate photographs might be left in his possession for him to recognize his mother, but photographs of the son could not be left in the mother's possession to serve the same purpose. Such blood relationships will surely not matter in heaven, but will we need introductions to almost every person we meet, including old friends and even our closest relatives?

And what of those with whom we may have corresponded for many years and yet have never seen, or may even have spoken to by telephone but still never met face-to-face? How shall we recognize them: or they us? Certainly not by the wearing of name-tags!

It was easy for the disciples to recognize the Lord, once their minds had accepted the reality of the resurrection, because of the marks on his body which He deliberately made use of to serve that very purpose. But it seems highly unlikely that any of the redeemed will come before Him with any such marks, with a limp or a missing arm or anything marring the perfection of his new body.

Thus we cannot allow ourselves the conceit of supposing that we shall automatically have the kind of magnificent body or beautiful body that would be appropriate in the prime of life, for such a body would not be appropriate in many cases as a means of identification. Some other principle of "identity" must be in view.

There is an identity which is non-photographic. Artists recognize this and are often guided by it even though to the uninitiated it looks like a distortion. Michelangelo, with his extraordinary skill as a sculptor, often portrayed his subjects in such a way as to make them both recognizable and unrecognizable: recognizable to those who already knew them with a measure of intimacy but unrecognizable to those who didn't. They were recognizable in that he had captured the soul of his subject, which would be familiar to those who knew: but unrecognizable to those who didn't know the soul of the subject because he disregarded to a great extent external appearance, leaving his portrait without adequate visual correspondence.

For instance, he portrayed Lorenzo the Magnificent who was his great benefactor, as a *soul* of great beauty (which he *was* in the eyes of many people) rather than as a somewhat mean-looking character such as we see on coins struck with his image during his lifetime. The coins showed what he *looked like*. Michelangelo showed what he *was*.

Again, if one examines his masterpiece, *The Pieta*, in which he portrayed the crucified Lord as it were "draped in death" across Mary's knees, one will notice that the face of Mary is if anything younger than that of the Lord. His purpose was undoubtedly to present Mary to the viewer as a soul of great beauty, apart from the fact that her attitude is one of sad resignation. But considering that she must have been by this time at least 50 years of age,* and considering that the Middle East is not "kind" to the faces of women as a rule, a photographic image might well fail to reveal the heart of Mary behind that ageing face. Michelangelo was not trying to be merely kind: he wished to be truthful, and unconsciously we read the truth.

We may well find that in our new bodies we shall all have a truer and deeper identity than mere photographic likeness. Perhaps the identity will not be established at all by shape or configuration. The shape or configuration that we shall each have may be precisely *the shape or configuration which will be a creation of the viewer in each case and not of the one viewed*. We shall see in our friend what will seem to us ideal: and yet another friend will see in that same person something quite different though equally satisfying his or her ideal. We shall not know we are looking at something different, we two viewers. And it will seem quite possible that when we look at ourselves, if there is ever such a thing as a heavenly mirror, we shall see what is ideal to us and it may be very different from what our viewers are seeing.

Could it not be that each of us will be recognized not by the normal visual impact of ordinary viewing but by the miracle of a transformation which will convert the visual signal in the mind of the viewer into an appropriate form that is entirely congruous with the personal nature of that individual as the viewer once knew him or her in this world?

*** By the time of the crucifixion, the Lord was approximately 33 years of age, and it seems rather unlikely that Mary would have been less than 17 years of age when He was born.**

The woman would appear as a woman to those who knew her as such, as a girl to others who had known her at an earlier stage, and conceivably even as a child to the mother who had died when that child was only a few years old.

The four Gospels present us with apparently different and intentionally contradictory accounts, though the central figure is clearly one and the same Person throughout. We have to assume therefore that the impression He generated was different to the minds of different viewers. Matthew saw Him as the Promised King, with all the earmarks of a kingly presence. Mark saw Him as a servant *par excellence*, which is quite another thing. Luke saw Him as a Man among men, as "the Son of Man" indeed. John saw Him not as Man with the attributes of God but as God with the attributes of Man.

He was all these things. He communicated his presence and his identity differently to different men, not because He was changeable in any way but because those whom God chose as writers of the inspired record each perceived Him according to their own personality and background.

I have no doubt that people from countries all around, regardless of their skin colour or characteristic appearance, saw the Lord as a true representative of mankind *in their own terms*. We have a slight intimation of this possibility in the way very small children will readily play with children of other races without any apparent awareness of difference in skin colour or features. It is possible therefore to be quite unaware of these things. Perhaps it is a mistake for us to portray the Lord "as one of us" with fair hair and blue eyes since people from other races are rightly offended by this form of ethnocentricity. The Lord was strictly the Son of Man, somehow escaping these formal limitations. I wonder what our illustrated Bibles and Christmas cards are doing, deeply stamped as they are with our image of what is becoming in man.

I believe that when we see the Lord, we, too, will each see Him differently. To the white man, He will be *our* ideal;

and so will He be to the black man his ideal; and to the Chinaman, and to the Eskimo, and indeed to "every man". Each of us has his or her own ideal of what is beautiful and what is formal perfection. He will meet all our ideals individually – not because He will be both black and white but because our resurrected minds will be so structured as *to filter our vision* appropriately.

And thus we shall somehow recognize beloved friends in the Lord no matter how changed in later life they or we may have been. Nor shall we need any introductions to those whom we never actually met – Adam, Job, David, Paul, Augustine, Luther, Whitefield .... We shall recognize by "essence" not by shape.

As He sees *us* in a perfect way that truly reveals the character of our redeemed spirit, so because we shall be like Him we shall see *Him* as He is; beautiful to behold, and surrounded by truly beautiful people, his redeemed children.

C. S. Lewis in a sermon titled "The Weight of Glory", made a perceptive observation which is particularly apropos of what we have been saying. He suggested that the most uninteresting and commonplace person one has ever spoken to may one day be such a creature that if one met him today one would be tempted to worship him. On the other hand, he may one day be such a creature, so full of horror and corruption, as to be conceivable only in a nightmare.

As he put it, "You have never talked to a *mere* mortal": by which I think he meant no man is just what we see before us, but rather what he may be potentially – for good or ill. The creatures with whom we joke, snub, exploit, or marry are potentially "immortal horrors or everlasting splendours".

It is a sobering thought: truly awful or truly wonderful. And it depends as much on the potential of the human body as it does on the potential of the human spirit. The concept of a chance evolutionary process producing a creature with such a potential for inexpressible beauty or unbelievable ugliness seems to me utterly absurd. We cannot account by such a means for either alternative – the horror of the effect of sin, or the glory of the effect of redemption. In this beauty or horror the body plays an essential role. That heaven or hell should be peopled with beautiful or horrible *ghosts* is really inconceivable, and the Bible certainly does not suggest such a thing. We are to be rewarded or judged "in the flesh". Our bodies are to share the glory or the shame.

Chapter 18

CONCLUSION: CAUSE AND EFFECT

"The true way in which to examine a spiritual movement is in its logical relations: logic is the great dynamic and the logical implications of any way of thinking are sooner or later certain to be worked out."

J. Gresham Machen

I am fully persuaded that an established fact is as sacred as a revealed truth. But is the *creation* of man "a revealed truth" or is it merely one possible interpretation of the Genesis account? And is human evolution an "established fact" or merely one possible way of interpreting the geological record? How do we decide? How does one distinguish interpretation of evidence from the truth itself? Is "truth", after all, not *reality* but only our interpretation, i.e., our *perception* of it? How are we to decide what is the correct interpretation of Genesis?

I have in my own library over 350 commentaries on Genesis, going back to some published in the 1600's. In addition I have the works of the Church Fathers during the first few centuries of the Christian era, among which are many treatises on the book of Genesis in various forms. Besides these, taking us even further back in time, I have some of the earliest commentaries and paraphrases of Genesis written by the Jewish people themselves. Probably no other book of the Bible has been subjected to such scrutiny, or studied so extensively, or argued about so assiduously. Why? Because *beginnings* are always important. If our Faith is a logical construct, here is its foundation. If the evolution of man is a truth, our foundation is in error. How can a Faith be *defended* if its foundation is destroyed?

Of course, we still have to interpret the sacred record: and of course, this in large part accounts for the diversities of opinion in these commentaries since each commentator brings his own pair of spectacles to the task. Yet apart from works written in the past one hundred years or so, these commentaries almost without exception have held to the view that man was created by a direct and immediate act of God, not merely in spirit but in body also. If long-held biblically informed opinion counts for anything, then either the modern accomodation to evolution cannot carry the same weight or for almost 2000 years the Church of God has been misled in a *basic* tenet of its faith.

This has indeed been the Faith of the Church for centuries, and so long as man's *origin* was understood in this light, his *destiny* was understood in the same light – as equally unique. Unlike the other animals whose bodies return to the earth and whose spirits are destined to do the same (Eccl.3:21 and 12:7), man's spirit goes upwards to God who gave it while his body awaits resurrection and reunion with it to reconstitute the individual in his entirety.

The destinies of the animal and the human body are diametrically opposed and it would be reasonable therefore to assume that their origins are equally diverse. The Christian evolutionist accounts for the origin of man's body by a wholly *natural* process but must then allow it to have an entirely super-natural destiny – which seems a strange incongruity. Can such an incongruity be defended? Can a body with such a super-natural destiny as man's body is designed for, have had the same origin that evolution demands for the animal body?

Evolution is also strictly a Faith because its presuppositions are, like all presuppositions, simply taken for granted. In the nature of the case, they are probably unproveable in strictly scientific terms. If man is merely an animal, then such a Faith entails a consonant view of his destiny. Each branch of science has its own canons of proof – in mathematics, logic: in astronomy, observation: and in the natural sciences, experiment. Because evolutionary theory claims to be a natural science, it ought to be subject to proof, but it isn't. It never can be, because what is past is past and cannot be repeated; and an essential of proof in any science is repeatability. We cannot step into the same river twice.

Nor can we prove the great articles of our Faith, for we cannot even prove the existence of our God! "He that cometh to God must *believe* that He is, and that He is the rewarder of them that diligently seek Him" (Heb.11:6). Faith is the key to understanding in such matters, not understanding the key to faith as we may prefer to think.

It was Anselm (c. 1033-1109) who wrote: "I believe in order that I may understand", and Abelard (c. 1079-1142) who, fifty years later, argued: "I seek to understand in order that I may believe". Anselm was right, because *every* world view starts with an act of faith and is intelligible only when that initial commitment has been made.

Every "system" of thought, including evolution, begins with a premise that is an act of faith. We have to allow this: and if we deny an opponent's premise or he denies ours, we have no starting point for meaningful dialogue. When we insist on proof of the basic premise, we bring to a halt all useful discussion. A premise cannot be proved, for then it would become a *conclusion* and cease to be a starting point.

Because the basic premise of the Christian Faith is completely opposed to the starting point of a strictly evolutionary world view, there is no real reconciliation possible except by abandoning any rigid adherence to logical extension. The Christian view allows forces and realities of a spiritual nature to be causal factors in the physical world. Evolution absolutely forbids them. The two positions are logically irreconcilable and thus it is hopeless to attempt a wedding of Christian theology and evolutionary theory. They are erected on different premises and can only be reconciled by the introduction of logical inconsistencies.

If man derives his body from an animal source with nothing added beyond what can be provided by nature, then he must be an animal and nothing more. And if he shares the origin and nature of animals, it is only reasonable to suppose that he will share their destiny. To argue that he *is* an animal but an animal with a soul introduced from outside the system, may be a sop for theology but it is quite unacceptable to the strict evolutionist. And even for theology such a concession is wholly inadequate because man's body is to share equally in the redemptive process. So that it is not enough to argue that a human soul is sufficient distinction: he must also have a body that is uniquely human.

Why is this not a sufficient concession to theology? Because the redemptive process required the vicarious sacrifice of a truly human body. If it should be asked in what way this redemptive process relates to the nature of Adam's body as created, the answer is to be found by considering the three following compelling affirmations which are locked together and must all be precisely honoured or none of them has any validity.

First of all, unless the body of Jesus Christ was by nature under no necessity of dying, it could not be sacrificed *vicariously*. Secondly, unless the body of Jesus Christ was truly *human*, the sacrifice of it was not vicarious *for man*. Finally, if His body *was* by nature free of any necessity of dying and was therefore indeed sacrificed vicariously for man, then Adam's body in its original form **must** also have been by nature *free of any necessity of dying.*

Such was Adam's body, and it cannot possibly have been derived from any primate body since all other primates are by nature subject to death. It must therefore have been a direct creation of God. *Immediate* creation thus meets all the theological requirements whereas *mediate* derivation by evolution meets none of them.

The Christian evolutionist takes refuge in the tacit assumption that it is the soul of man which is unique and this is all that matters. As for man's body, since it is animal in origin, it cannot have any transcendental significance. In short, the body is not essential to man's continuance in eternity. In this view, man is man only because of his soul. The biblical view of man as a unique body/spirit entity has to be abandoned. Man can exist just as well as a mere ghost, even if that ghost supposedly has to assume the *shape* of his body for purposes of identity!

To the evolutionists, there can be nothing in nature of which nature is not the author. No discontinuities in the great chain of being from amoeba to man can be allowed. No divine intervention, no miracle to upset the strict chain of physical causality, no revelation to point up the inadequacy of the human mind in its interpretation of the evidence.

The Christian who admits to faith in either creation *or* revelation is suspect, for evolution is a strictly deterministic, materialistic, non-theistic philosophy. Thus it is difficult to maintain a position that falls *between* the two diametrically opposed world views. The passage from a little admission of doubt to a frank abandonment of any biblical faith at all is along a slippery path, and those who start upon it innocently enough, soon find it more and more costly to stop the slide. Or if they do, they will in the end find it increasingly difficult to defend what faith they retain except on an emotional basis.

This abandonment of rational defence is made a little easier if one really has no understanding of why the issue of the *body* of man is so important; but such ignorance of the true situation hardly encourages a strong and mature Christian faith. Unfortunately, while we have a highly developed theology of the spirit, there is as yet no convincing parallel theology of the body. The importance of the body has been sadly neglected.

To a large extent our problem is the result of a paradox. In spite of a gross materialism which infects us all, we have somehow overlooked the importance of a material body, even though the New Testament lays such an emphasis upon it. We have tended to settle for the Greek view that it is a prison to be abandoned as soon as possible. It thus becomes easy to view its evolutionary origin as quite acceptable, for in this view it is only a *temporary* part of our being in any case – not something we shall have throughout eternity. Where it came from is as irrelevant as where it is going to. Whereas it is in fact part and parcel of our hope of glory!

We betray this hope of glory by barely perceptible stages, yielding on some supposedly "innocuous" point (such as the vast antiquity of man) which then becomes a first step down the slippery decline. And we do it in a way that causes us little concern because the element of betrayal of our Faith is not apparent. We just don't see that it matters.

Soon it becomes necessary to go a step further and abandon the idea of a real Adam and a real Eve such as we find in the biblical record – and with it goes the story of the Fall. As their concrete reality slowly recedes, we are easily led to make the tacit assumption that if there *was* an Adam and an Eve in such a distant past, they would obviously have looked much like the ape-men of current reconstructions, and very soon the derivation of Eve out of Adam becomes clearly mythological.

The next step is comparatively painless. Some kind of ape is in fact *in our direct line of descent*. Since the further back we go by natural generation the more "primitive" we assume we were, early man is soon indistinguishable from the rest of the primates, so "why not simply join the family"?

By now our passage down the slope is going full-tilt. It

is quite exhilarating. We are coasting freely with the crowd. We no longer stand alone: and the comforting thing is that it doesn't seem to matter. We don't *appear* to have sacrificed any essential part of our Christian Faith. Yet somehow the "old Book" seems to have lost its power to inspire or rebuke or encourage. And we come across the references to the Lord Jesus Christ as the Last Adam and scarcely realize that the term is now meaningless because the First Adam has been lost in the mists of evolutionary antiquity two or three millions of years ago and cannot possibly have truly been a prototype of the magnificent figure we meet in the Gospels as the Last Adam.

But it is too late to retract. At least, the cost will be so great in terms of "reputation for sanity" and perhaps even job security, that we cannot contemplate it. And so we settle in the end for a philosophy that is totally inconsistent with the Faith we once lived by. To do otherwise requires more and more moral courage, while the exhilaration of apparent "progress' is hard to give up. Without a strong conviction that one's course is wrong, there is no incentive to try to reverse it. And one cannot have any strong conviction in the matter unless one understands something of the reason for ever having had such a conviction in the first place.

No one really wants to be the only man in step. It takes a lot of stamina to go against the current. It is difficult to sustain a position which is only reasonable to those who allow its premises, and the enemy is adamant in rejecting these premises.

But the logic of the Christian Faith has such an inner consistency that it cannot be tampered with or adjusted without virtually rendering it irrational and therefore difficult to defend. To surrender any part is either to surrender the whole (as many have done in recent years) or to retain what is retained in a way that is incomprehensible to those who have *not* made the surrender.

Such inner inconsistency may allow a certain level of

continued fellowship along purely spiritual lines but it certainly destroys fellowship of *mind* which is always such a delight among the Lord's people. The sad thing is that such forms of disagreement are often the most distressing because theological incompatibility seems more conducive to disharmony than spiritual incompatibility. Mental convictions are more intractible: we generally find it much easier to bend on purely spiritual matters than on theological ones. Heart is more "forgiving" than mind. Protestant charismatics who attach little importance to theology can have close fellowship with Roman Catholic charismatics. It is when either party becomes theologically adept that their fellowship is endangered.

We will not repeat what we have said in the first chapter regarding the shift in Evangelical Theology which took place at Princeton Theological Seminary immediately after the publication of Darwin's *Origin of Species* in 1859. But it may be worthwhile to pause for a moment to reflect upon the first step taken on this slide, because it seemed at the time so innocuous.

Sir Charles Lyell (1797-1875), one of the first British geologists, published his *Principles of Geology* between 1830 and 1833. In this three volume work he established that a very long period of geological history must have *preceded* the introduction of Adam while the earth was being prepared for him. He did not seem to have any doubts about the reality of the Genesis account of what happened to Adam and Eve and their immediate descendants, only that the six days which preceded Adam's creation had to be interpreted as a far greater period. But in a later work, *The Antiquity of Man* published in 1863, Lyell argued that Adam must himself have appeared on the scene far earlier than the 4000 BC of biblical chronology because his remains were being found increasingly in rocks of great age.

As we have seen, William Green accomodated the Bible

to geology in this matter by proposing that no absolute biblical chronology was possible due to gaps in the genealogical records. What neither Charles nor Alexander Hodge recognized in Green's thesis was the fact that the argument for the existence of gaps is entirely circular. If it is asked, "How do you know there are gaps?", the answer has to be, "Because parallel records elsewhere in Scripture show them up by supplying the missing names!" How else would one know there was anything missing?

But if the gaps are filled in elsewhere in the same book, there *are* no gaps! In other words, we know that in some instances the genealogies of the Bible *are* abbreviated, but we only know this because the abbreviation is pointed out to us elsewhere *by supplying what is missing.* In short, nothing is actually missing; it is only that the data are presented elsewhere. Were they not to be found "elsewhere" we could have no knowledge of "the gaps" in what preceded. The *only* way we can become aware of the abbreviations in the genealogies is by being warned later on that such abbreviations exist because we are given the names of the individuals who must be inserted therein to make the record complete. When we are not given these names we have no right to assume there are still gaps. Thus, in effect, the genealogies do not actually contain unfilled gaps at all. The only concrete evidence for a gap is the "filling" supplied elsewhere, which only *reveals* the gap by *closing* it!

If, for instance, a piece of information in a detective story is given later in the book, would a reviewer say, "Unfortunately, the story has gaps in it"? The identity of the murderer may be left by an author until the very last sentence, but this does not mean that the murderer's identity is missing! Omissions in *one* place supplied in another place are not really gaps at all, but only temporary withholdings according to the author's intention.

If on an examination paper, for the sake of neatness and ease of reading, one were to work out a mathematical sum but omit some of the commonplace steps in the calculation, and then append these omitted calculations on a final sheet of paper, it would be quite improper for the examiner to argue that the student has actually left gaps in the calculation.

In the Bible these so-called "gaps" are made manifest and clearly indicated by what is recorded elsewhere in Scripture itself. There is no *evidence* of any other gaps. How could there be? What would be the nature of the biblical evidence of any such other gaps which are not elsewhere made manifest? How could one identify such a gap at all unless it was elsewhere "filled in" and by that very means removed? William Green was quite unjustified in claiming there are still actually gaps in the genealogies without providing literary evidence for the assertion.

Furthermore, if a vast antiquity far beyond the 4000 BC traditional date is demanded, there are other ways in which a great antiquity for the world prior to the creation of man can be allowed for. For example, the days of Genesis might be viewed as days on which revelation was given to Moses; or they might be taken to mean ages; or we may introduce a hiatus between Genesis 1:1 and 1:2, and so on. One may disagree with these alternatives, but they have all been proposed by biblical scholars over the centuries and are still held by many serious and well-informed students of Scripture.

But there is no way, if we limit ourselves strictly to biblical evidence, that Adam as the first man of whom the Last Adam was a true representative, can be pushed back into a vast antiquity as currently demanded. There *are* ways of accomodating long periods *prior* to Adam, but not of pushing Adam himself back into a dim period a million or more years ago.

That Adam was indeed the first man seems clear from the Bible. To begin with, there was prior to Adam *no man* to till the earth (Gen.2:5): and when Adam was created to perform this function he was strictly *alone* in the world

(Gen.2:18). In his aloneness there was *no helpmeet* for him, and one had to be formed out of his own flesh (Gen.2:20). This helpmeet then became the mother of *all* living, i.e., of *all* other human beings (Gen.3:20), none excepted. And to clinch the matter, 1 Corinthians 15:45 tells that Adam was truly "the *first* man".

Suppose for the sake of argument that this first man really was introduced upon the scene at least a million years ago, how do we fit this million-year silence as to his subsequent history into the setting of the first three chapters of Genesis? Was there "so long ago" a Garden of Eden in which a helpmeet was brought to him, formed out of his own body? And were these two then tested in the matter of a fruit they were not to eat?

And did the disastrous results of their disobedience then affect the whole human race subsequently so that every individual in Adam's line *throughout all those unrecorded millenia* was a fallen creature about whom Scripture is totally silent? One has to bear in mind that the Last Adam stood as a potential substitute for *all* the descendants of the First Adam – and this would therefore include all the human beings living in the world throughout these silent millions of years. *All* of them, no matter how ape-like or how little like the Last Adam they might be, would still have to be such that He was nevertheless truly in the same image: and these untold millions of intermediate forms would still have to be truly represented by this Last Adam both physically and spiritually or they would not be "redeemable" under the terms that clearly apply according to the biblical strategy of Redemption.

Shortly after the expulsion from the Garden, this first man and his wife consummated their marriage and bore two sons who grew up, one to become a farmer and the other a herder. We actually have their names! One of these sons in a fit of jealousy killed his brother, fled from the company of his other brothers and sisters and, taking his sister-wife with him, built the first "city", naming it after his own son Enoch. All this happened within two generations of Adam's first appearance.

And there followed during the next three or four generations the development of a high civilization which included all the arts and technologies we have already mentioned. Suddenly without any evidence of a hiatus of even a few centuries, let alone a million years, we find ourselves on the very threshhold of an advanced culture abundantly witnessed by the findings of archaeology.

Bringing Adam into the historical period makes sense but what does one do with a hiatus of almost a million years? I remember a brief article on the evolution of man which occurred in a well-known Christian magazine as part of a series by various authors on evolution generally. It was written by a Christian anthropologist about 27 years ago. In his article the writer said, in effect, that it was now reasonably certain that man was at least 100,000 years old. He was clearly quite willing to adjust his thinking about Genesis to suit this new position. One wonders how he felt, when fifteen years later, 100,000 years had become an even more confident "million" years: and now we are being told that our roots probably go back 3 million years. *At what point* does one say, "Hold! Enough." And having done this, what position does one take? At what point would the writer of that article now say, "Here is where Adam was introduced"?

This kind of open-ended stance can soon become so nebulous as to leave one without any position at all, or to put the matter slightly differently, we would really have to say that there was no Adam, ever. The *first* Adam is thus eliminated: what does this do to the Last Adam?

Somewhere we have to find room for these millions of years, and if we do it is obvious that the very simple and straightforward record in Genesis of what happened to Adam and Eve (and the whole human race in so far as they were latent in them) becomes a shambles. The foundation of the redemptive history of man is utterly destroyed. And it seems that the great Princeton defenders of the Faith felt no need to attempt any reconciliation between these two historical perspectives, nor were they aware that by allowing the evolution of Adam's body they were placing in jeopardy the theology of Redemption they otherwise so ably expounded.

Perhaps they surrendered too readily. The evolutionary picture itself is tending slightly towards a position regarding the origin of the human species which lies a little closer to the biblical concept of direct creation, though this would certainly not be admitted by those who advocate these newer concepts of evolutionary process.

As we have seen (page 37), Goldschmidt's "saltations", then Simpson's "quantum leaps", and now Gould's "punctuated equilibrium", are all attempts to account for the sudden appearance of entirely new species, of which man is but one. But they are descriptive terms, not explanatory. Yet they *are* at least admissions of a fact that is surely most simply explained as evidence of creative intervention. The biological record is by no means a smooth "gradualist" one but is marked by abrupt changes from time to time involving sudden extinctions on a mass scale and sudden new emergings, leaving many gaps that have persisted despite continued assurances that they will be found. These gaps are almost universally admitted, as is also the fact that they can no longer be attributed (as they were formerly) to the "imperfection of the fossil record".

Thus while the *suddenness* of man's appearance is not perhaps the problem that it was, the assumed *timing* of his appearance certainly remains a difficulty for anyone whose faith in Scripture stands firm. And I do not think that the biblical account can ever be made to accomodate the antiquity that is still being demanded. Personally, I am convinced that the arguments for this vast antiquity will in due course be modified by fresh evidence and the Bible vindicated, as it always has been. Certainly, within the *historical* period, archaeology has consistently verified the biblical record, not simply in a *general* way but in a way that confirms it in its most literal reading. Historically it has stood every test – and I am convinced that the story of Adam and Eve is part of history, not prehistory.

In increasing numbers there are devout Christians who nevertheless take the evolution of man's body for granted. They tend to regard Genesis as mythology or poetry or allegory – but they have not adequately considered what such an approach does to the congruity which is so strongly reinforced in the New Testament between the First and the Last Adam.

The bond between the only two truly *human* beings, both of whom were thus called Adam, is entirely predicated on a miraculous origin in *both* cases: the creation of the first man Adam, which was clearly a supernatural event; and the virgin conception of the Last Adam, which was also clearly a supernatural event.

A body of *animal* origin acquired by evolutionary processes is an entirely different thing from a body of *divine* origin acquired by direct creation. As to the former, it is clear that such a body must by nature be subject to death, the ancestral line being through some primate channel where death is natural. As to the latter, such a body becomes subject to death not by nature *but only as a penalty*.

The whole Plan of Redemption hinges upon this difference because the Last Adam cannot by *nature* be subject to death and still make a truly vicarious sacrifice of Himself. He would merely be paying a debt to nature before the expected time.

We have to recognize that whereas an animal dies NATURALLY because it is necessary, a man dies UN-NATURALLY because it is a penalty – he is, in fact, executed. By contrast, the Lord Jesus died SUPER-NATURALLY because He died vicariously, himself alone being the executer. As soon as this important truth is lost sight of, we have lost the rationale of the divine plan for man's Redemption.

The origin and the death of the First Adam were not *natural* events: neither were the birth and death of the Last Adam. These four events stand or fall together. To speak of the latter but deny the former is to render meaningless the terms *First* and *Last* upon which so much hinges in the Plan of Salvation.

We need not only a "theology of the spirit" but also a "theology of the body". In every Systematic Theology there is a section called *Anthropology* which is intended to treat of the nature of man. But it is remarkable how little is said about the importance of man's body not only in this life but in the world to come. After all, this *is* an absolutely essential component of his nature. At any rate, while we have a plethora of Christian works on the importance of man's spirit, they tend to eclipse the importance of man's body. It is really no wonder that the evolution of man's body has been accepted so readily and so widely.* There has seemed no compelling reason *not* to accept it and the pressure to do so is very great indeed. If one can see no reason from the point of view of the validity of one's faith as a Christian, why *not* accept it when everyone else appears to be doing so?

Moreover, the theory of evolution has firmly established itself as a theory with tremendous heuristic potential, a key to everything that involves change and development – or so it seems. Since it is a recognized principle that a "useful" theory is not overthrown merely by contrary evidence but only by a better theory, and since the "better" theory that we prefer as Christians is predicated on certain metaphysical premises that are totally unacceptable to the scientific com-

---

* Even the Roman Catholic theologians, conservative as one expects them to be, have now been given permission to adopt evolution as a working hypothesis, though they too will find it introduces unresolvable contradictions into the theological rationale of the Atonement. This unfortunate step was given approval in 1951 by Pope Pius XII in an encyclical titled *Humani generis*, which allows evolution to be applied to man's body but not to his soul.

munity at the present time, it looks as though we really have no alternative that can hope to succeed in opposition to it.

It is fundamentally the premises that are at issue, and here faith is the deciding factor as to which premise is chosen. And faith in the reality of a Creator is not achieved by rational argument but only by a leap in the dark, a sudden change of viewpoint. It would therefore seem that our only hope in this warfare lies in a mighty revival among ourselves and a gracious infusion of saving faith among those who are our chief opponents. Perhaps we need to pray more for the salvation of our opponents and not just try to marshall more facts – though we certainly should not enter the fray unarmed!

In spite of the fact that our warfare in this matter is a spiritual one rather than an intellectual exercise or a battle of wits, it is still important to keep clearly in mind that the use of reason cannot be abrogated. To communicate with an intelligent person we must be able to give a *reason* for the hope that is within us. Irrationality is not necessarily evidence of great faith. But the relation between the direct creation of Adam's body and our "hope" must at least be clear to *us* if our words are to carry any weight.

It is my conviction after watching this warfare between evolution and creation for over half a century that there are indeed *fatal* flaws in the theory of evolution that ought certainly to be kept before the general public. On the other hand, there are also theological consequences of which the Christian public has for too long been almost wholly ignorant.

But theological arguments for rejecting the evolution of man's body cannot be used as so much 'ammunition' in this warfare because they are not based on the same premises upon which our opponents base their case and they therefore carry little or no weight with them. Nevertheless these arguments (a number have been examined in this book) *can* and *should* be used to arm the Lord's people against a too ready capitulation to evolution as it applies to man's body, even if we are not adequately equipped to challenge all the biological data on which the theory is based.

But above all, we must bear in mind that whatever we allow that diminishes the reality and the historical significance of the First Adam inevitably reflects upon the reality and historical significance of the Last Adam. That is why, in my view, the issue is so very serious and why it must be fought on both fronts, the biological *and the theological.*

Why should we contest the evolution of man's body so vehemently? Because half of our humanity is wrapped up in it. It is a body which was glorified by the Incarnation, a body which has made us redeemable creatures, a body which has a future *beyond* death and will be essential to our recognition of one another throughout eternity.

The New Testament asserts that this body will be resurrected in identifiable form and placed beyond the power of death for ever. Moreoever, this promise is applied to Adam himself and effectively to every one of his descendants, how ever far back we place him in time.

This prospect demands a unique origin because the redemptive process by which that destiny is to be effected hinges upon the fact that man began in the first place with a body *which was potentially immortal.* Such a body cannot have arisen by any evolutionary process, because it was designed for a unique destiny not promised to any other creature.

Evolution is confessed by its own chief advocates to be a philosophy without any such hope. We believe that by the super-natural birth and death and bodily resurrection of the Last Adam, of whom the First Adam was the prototype, we can have not only a bright hope for the future, but a *certainty* of it.

INDEX OF BIBLICAL REFERENCES

INDEX OF NAMES

INDEX OF SUBJECTS

Printed and bound in Canada by John Deyell Company